Praise for

The Casebook of Forensic Detection

"This well-organized compendium covers cases . . . arranged according to the methodology by which they were solved. . . . ranging from ballistics through DNA typing, fingerprinting, odontology, serology, and toxicology. . . . even the most dedicated devotee of the genre will find much that is new in these exciting accounts of the brilliant and persistent scientific work that brought murderers like John List, Ted Bundy, and Jeffrey MacDonald to justice." —*Publishers Weekly*

"Contains ample material to hold the attention and foster interest in science." —*Science Teacher*

"A mystery novelist's essential resource guide." —Book News, Inc.

"Pithy, concise, and remarkably accurate." —*Science Books & Films*

Some of the cases covered:

- Computerized fingerprint analysis stops one of California's worst killers—the Nightstalker.
- Forensic sculpture reconstructs a face from a skull and catches the man in the clown's suit—John Wayne Gacy.
- Computer-enhanced imagery helps capture one of *America's Most Wanted*—John List.
- Crime solving changes forever—psychological profiling and New York City's "Mad Bomber."
- DNA typing and one of twentieth century's great mysteries—was Anna Anderson the real Anastasia, the daughter of Czar Nicholas II, and the last of the Romanovs?
- How forensic experts identified a victim with only one one-thousandth of her body parts.
- A psychological profile that helped catch a dangerous criminal who drank his victim's blood.
- What bones reveal about the victim—and the
- Computer enhancement catches a killer

Most Berkley Books are available at special quantity discounts for bulk purchases for sales promotions, premiums, fund-raising, or educational use. Special books, or book excerpts, can also be created to fit specific needs.

For details, write: Special Markets, The Berkley Publishing Group, 375 Hudson Street, New York, New York 10014.

THE CASEBOOK OF
Forensic Detection

How Science Solved 100 of the World's Most Baffling Crimes

Colin Evans

BERKLEY BOOKS, NEW YORK

THE BERKLEY PUBLISHING GROUP
Published by the Penguin Group
Penguin Group (USA) Inc.
375 Hudson Street, New York, New York 10014, USA
Penguin Group (Canada), 90 Eglinton Avenue East, Suite 700, Toronto, Ontario M4P 2Y3, Canada
(a division of Pearson Penguin Canada Inc.)
Penguin Books Ltd., 80 Strand, London WC2R 0RL, England
Penguin Group Ireland, 25 St. Stephen's Green, Dublin 2, Ireland (a division of Penguin Books Ltd.)
Penguin Group (Australia), 250 Camberwell Road, Camberwell, Victoria 3124, Australia
(a division of Pearson Australia Group Pty. Ltd.)
Penguin Books India Pvt. Ltd., 11 Community Centre, Panchsheel Park, New Delhi—110 017, India
Penguin Group (NZ), 67 Apollo Drive, Rosedale, North Shore 0745, Auckland, New Zealand
(a division of Pearson New Zealand Ltd.)
Penguin Books (South Africa) (Pty.) Ltd., 24 Sturdee Avenue, Rosebank, Johannesburg 2196,
South Africa

Penguin Books Ltd., Registered Offices: 80 Strand, London WC2R 0RL, England

Interior text design by Kristin del Rosario.

PRINTING HISTORY
John Wiley & Sons trade paperback edition / October 1998
Berkley updated trade paperback edition / August 2007

Library of Congress Cataloging-in-Publication Data

Evans, Colin, 1948–
The casebook of forensic detection : how science solved 100 of the world's most baffling crimes / Colin Evans.
p. cm.
Includes index.
ISBN-13: 978-0-425-21559-3
1. Medical jurisprudence—Case studies. 2. Criminal investigation—Case studies. I. Title.
RA1053.E93 2007
614'.1—dc22

2007015095

PRINTED IN THE UNITED STATES OF AMERICA

10 9 8 7 6 5 4 3 2 1

"Freed"—
this one's for you.

Acknowledgments

In any book of this kind the role of the writer is but a contributory part; most of the credit is due to the selfless support I received on both sides of the Atlantic. Among the individuals who gave so graciously of their time and effort are William H. Garvie, FBI; Brad Smith, Los Angeles DA's office; Amy Louise Kazmin, *Los Angeles Times;* Wayne T. Seay, Chief of Detectives, Nassau County Police Department; Tina Vicini, Chicago Police Department; and Gerard J. Sciaraffa, Circuit Court of Cook County, Illinois.

Research staff at the New York Public Library, the National Archives, the Library of Congress, the Broward County Library in Florida, and the British Newspaper Library at Colindale, London, all provided unstinting and selfless assistance.

As always, David Andersen shared his encyclopedic knowledge of crime and his enviable book collection. Both are greatly appreciated. Greg Manning, too, helped ease the burden. Samantha Mandor at Berkley Publishing has been unfailingly supportive. Thanks also to Mary Ray Worley at Impressions Book and Journal Services, whose meticulous copyediting helped clarify so much in the manuscript. I owe a special debt of gratitude to my agent, Ed Knappman, who first suggested this volume and who offered many constructive ideas in its shaping. Needless to say, while all of these people contributed so much to this book, responsibility for any errors is mine alone. Finally, a very special thank-you to Norma, without whom none of this would have happened.

Contents

INTRODUCTION

If history decides to remember the twentieth century as the century of science, then it will also record that a sizable portion of that science was devoted to catching criminals. Although rudimentary attempts at scientific detection date from the mid-1700s, only in the past one hundred years has crime fighting been put on a truly systematic footing. The results have been remarkable—enough to astound even Sherlock Holmes, the fictional forerunner of today's technosleuth. But whereas Sir Arthur Conan Doyle's creation combed crime scenes with a magnifying glass, his modern counterparts employ ultraviolet light, lasers, DNA, spectrographs, neutron activation analysis, computer software, scanning electron microscopes, blood grouping, and any number of other tools to identify that which we can't see and analyze that which we can.

The Casebook of Forensic Detection is an attempt to chart the progress of scientific detection through concise and accurate summaries of one hundred criminal cases from around the globe, all of which played their part in establishing the role of science in modern crime fighting. Selection has largely been dictated by a desire to inform and entertain and will be of use to both the casual reader and the serious researcher. Although some cases have become medicolegal classics, others are barely known. But of one thing you can be sure: None is ordinary. Each has some facet, some twist, that elevates it from the pack. Inevitably, space constraints have made it necessary to exclude many deserving cases. I hope the reader will agree that all of those chosen fully merit their place, and if your own particular favorite should happen to be absent, then please accept my apology. A bias toward American and European cases is unavoidable, because both regions have figured so prominently in the forensic crusade. But nonetheless, the major contributions of Australasia and South America have not been overlooked.

For ease of reference, the various topics, such as Ballistics, DNA Typing, Fingerprinting, and Trace Evidence, are listed in alphabetical order. Then, the individual cases within each category are arranged chronologically. In this way, a clear picture of how each specialty has evolved can be formed.

Populating these stories, beyond the victims and villains, are the forensic pioneers. Some were brilliant, carving landmarks in the annals of forensic medicine; others advanced opinions with an arrogance that only experience would subdue. For the most part, early juries regarded them all as demigods; nowadays the average juror is more circumspect, demanding ever higher standards of accuracy. The FBI Laboratory in Washington, D.C., and its British counterpart, the regional Home Office Forensic Laboratories, have sought to provide this probity. Their influence has been felt around the globe. Now virtually every country in the developed world has analytical facilities that were once the stuff of science fiction.

But although Sherlock Holmes would have been flabbergasted by such quantum leaps, at least one of his principles remains intact: observation. Without it, even the most sophisticated technology is not worth a jot. Notice within these pages how often hunches or intuition have played a part, how often something struck the investigator as being not quite right. Later in the investigation, science would confirm these initial suspicions; without them, science might never have been employed. Sadly, as forces and resources wilt under burgeoning crime rates, human observation necessarily becomes less of a factor. The caseload is now so great that crimes have to be prioritized, which in turn throws a greater burden on technological resources. However, if the history of forensic science is anything to go by, then crime laboratories everywhere will rise to the challenge. Anything else is unthinkable.

BALLISTICS

Strictly speaking, ballistics is the science that deals with the motion of projectiles and the conditions that affect that motion, but in criminal investigation it has come to mean specifically the study of firearms and bullets. Firearms have been with us for several centuries—the first handguns were used by Arabs around A.D. 1200—and as early as the sixteenth century, engineers realized that etching a spiral groove into the gun barrel imparts spin to the projectile, giving it much greater accuracy. This groove, or *rifling,* leaves the distinctive marks known as striations on the bullet itself and forms the bedrock of modern ballistics study.

To the expert, variations in bullets fired from different guns are immediately apparent. It is possible to determine that a bullet has been fired from a particular weapon primarily because of how the barrel is manufactured. First, it is smooth-bored, then reamed to specification diameter before being rifled. Because the tools used to make a barrel wear down minutely with each succeeding gun, it is impossible to produce two identically rifled barrels. Bullets fired from different guns always have different striations. By the same token, each barrel retains its own unique characteristics, which it imparts to every bullet it fires.

The barrel is not alone in leaving identifiable traces on a cartridge; other gun components may also leave marks. When fired, the bullet is driven forward through the barrel; simultaneously, its shell casing hurtles back against the breech face. Any imperfection on that breech face impresses itself on the case head. The firing pin, the extractor, and the ejector post may each etch marks on the head or shell casing.

Another factor to be considered is impact. A bullet that strikes human bone or any hard object can have its shape radically altered. For this reason, it is often more reliable to compare test firings with casings taken from a crime scene, if available, rather than the bullet itself.

The modern cartridge was invented in France in 1835 and consists of a casing with a soft metal cap holding the primer charge. When struck by the gun's firing pin, the primer ignites the main propellant charge, expelling the bullet from the gun and leaving the case behind. With the invention of smokeless powder at the beginning of the twentieth century came the need for a stronger bullet. The new powder's greater propellant velocity meant that earlier lead bullets were too soft to be gripped by the rifling in the barrel and tended to get stripped, fouling the barrel. This led to the introduction of the metal-jacketed bullet, usually made of cupronickel.

Whereas the old black powder left distinctive marks on the hands of the firer and around the wound (if fired within a close enough range), modern smokeless powder leaves traces that can be detected only through chemical testing. At one time suspects' hands were examined to see if they had fired a weapon recently. The test was designed to detect the presence of nitrates, but because nitrates are prevalent in so many innocuous household products, this practice has largely been abandoned.

Identifying the guns that bullets are fired from used to be a laborious business. Computers have changed all that. DRUGFIRE, run by the Federal Bureau of Investigation (FBI), is an automated computer program that links firearms evidence from different serial shooting investigations. Digital images of bullets from crimes fed into computers can be accessed instantly by any law enforcement agency that is hooked up to the system. DRUGFIRE is installed in more than one hundred crime labs worldwide, where it has processed more than 65,000 cases. Other computer programs work along similar lines. If a possible match is found during the screening process, it is still only a guideline; the final determination is made on a comparison microscope (see the Charles Stielow case on page 5).

Charles Stielow

DATE: 1915

LOCATION: West Shelby, New York

SIGNIFICANCE: The role of expert witnesses has often been shrouded in controversy. For the most part their evidence is given impartially, but, as demonstrated here, arrogance and greed often cloud purportedly objective testimony.

At around 6 A.M. on March 22, 1915, Charles Stielow, a German immigrant of enormous strength and feeble intellect, awoke at the farm where he worked in West Shelby, New York. Walking to the barn to begin his day's work, he discovered the body of Margaret Wolcott, the farm housekeeper. She had been shot to death. A trail of blood led Stielow to the farmhouse, where he found his employer, Charles Phelps, unconscious on the floor. Although still alive, Phelps would not survive the morning. Autopsies revealed that both victims had died from .22 caliber gunshot wounds. Because the farmhouse had been ransacked and money had been stolen, police ascribed the murders to a bungled robbery.

At an inquest on March 26, Stielow, whose grasp on events was childlike at best, denied owning any .22 caliber weapons. But his brother-in-law, Nelson Green, who was if anything even more retarded than Stielow, admitted to police that he had hidden no fewer than three weapons—all .22 caliber and all belonging to Stielow. He claimed to have acted on his brother-in-law's instructions. After a brutal interrogation, Green confessed that he and Stielow had committed the murders.

Stielow was arrested and held at the local jail for two days without either food or sleep. After being questioned around the clock, he admitted to owning the three weapons but said he had told Green to hide them only because he knew that a search was on for all .22 caliber firearms. Promised that if he confessed to the murders he would be allowed to go home to his wife, Stielow did so. However, despite making the confession, Stielow resolutely refused to sign any statement to this effect.

"Expert" Witness

At this point, one of the shadiest characters in the history of American jurisprudence entered the proceedings. Using the title of doctor—a qualification that existed only in his own fertile imagination—Albert Hamilton had been hoodwinking juries for years, most often on behalf of unscrupulous prosecutors and always at considerable profit to himself. A shameless self-promoter, Hamilton exuded an air of immense authority on a wide variety of subjects. Unfortunately for Stielow and Green, one of those pseudospecialties was firearms. Hamilton had skimmed a few European tracts on early attempts at bullet identification and had equipped himself with a microscope and camera—sufficient, he reasoned, to add the appellation "ballistics expert" to his résumé.

If Hamilton was an expert in anything at all, it was amateur psychology; he knew just how impressionable juries of that era could be when bombarded with overblown jargon and intimidating photographic enlargements. After inspecting Stielow's revolver, he claimed to see an "abnormal scratch" in the barrel—the same scratch, he asserted, that could be found on the bullets removed from Phelps's body. To bolster this claim, Hamilton shot several photographs of the scratch.

As a result, Stielow stood trial for his life. During the trial, the judge was plainly troubled by the circumstances of the confession and made his skepticism known to the jury. He was also concerned about the apparent lack of motive; no trace of the missing money had been found in Stielow's possession. To counter, of course, the prosecution had the testimony of "Dr." Hamilton. He did not disappoint. Brandishing his photographs, Hamilton loftily declared that the murder bullets had come from Stielow's gun and no other.

All of this puzzled Stielow's counsel, David A. White. He examined the photographs carefully and professed himself unable to spot the telltale scratches on the bullets. Hamilton, with breathtaking temerity, glibly explained that by mistake, the photographs showed the side of the bullets opposite the scratches. So authoritative was Hamilton's reply that this astounding response passed without challenge.

And there was worse to come. When White asked Hamilton to point out the alleged scratch in the revolver's barrel, Hamilton replied that this was not possible. Because the cartridge fitted the barrel so tightly, he said, the explosive gases had nowhere to go but forward. Consequently the bullet had acquired so much momentum that its lead content had expanded at the muzzle and filled in the scratch, thus rendering it invisible!

Today it is easy to dismiss such nonsense, but in 1915, with firearms and ballistic identification still very much in its infancy, Hamilton's testimony was powerfully compelling. The jury certainly thought so, and Stielow was condemned to death. (Earlier, Nelson Green had pleaded guilty and received life imprisonment.)

On death row, Stielow came to the attention of Spencer Miller, Sing Sing Prison's deputy warden. In conversations with the simple-minded laborer, he became convinced of Stielow's innocence and referred the case to a group of New York women, known as the Humanitarian Cult, who were dedicated to the abolition of capital punishment. Their efforts on Stielow's behalf were tireless and at times nerve-racking. Three appointments with the executioner were halted only at the final hour; the closest call was in 1916, when Stielow came within forty minutes of being strapped into the electric chair before a stay of execution was phoned through.

Other Suspects

Investigators hired by the Humanitarian Cult learned that two drifters named Erwin King and Clarence O'Connell, then serving sentences for robbery, were known to have been in West Shelby on the night of the murder and had been heard discussing the crime the next morning—before it was public knowledge. Urged by the women to clear his conscience, King finally confessed that he and O'Connell had committed the dual murder.

This revelation did not please the Stielow prosecution team, who whisked King from prison and grilled him for several days, after which he dutifully recanted his confession. But it was too late. News had already reached Governor Charles Whitman in Albany and a commission was appointed to investigate the circumstances of Stielow's conviction.

The commission was headed by George Bond, a Syracuse lawyer who chose as his assistant Deputy Attorney General Charles E. Waite. After questioning the principals, both men were convinced that King and O'Connell were guilty.

However, there was still Hamilton's evidence to overcome. An officer examined Stielow's revolver and immediately formed the opinion from the rust and grime buildup in the barrel that the gun had not been fired for several years, certainly not since well before the night of the murder. Also, Hamilton's pronouncement that the barrel's tightness did not permit the backward expulsion of gases was shown to be nonsense when the officer held a piece of paper behind the gun and pulled the trigger. A sheet of flame flew back, igniting the paper instantly.

Two sets of bullets were sent to the Bausch and Lomb Company in Rochester for microscopic analysis by Max Poser, a specialist in optics. Even under the highest magnification, Poser was unable to locate the alleged scratches described by Hamilton on either the test bullets or the murder bullets.

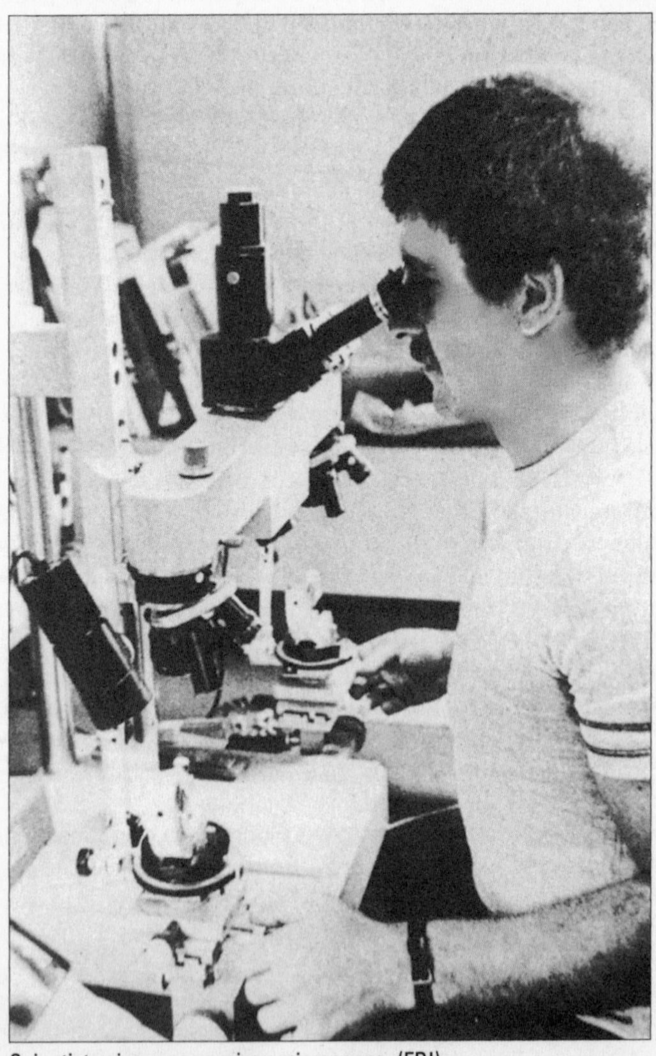

Scientist using a comparison microscope. (FBI)

All of this, of course, could be taken as merely one scientist's opinion against another's, but then Poser examined Stielow's revolver and made a discovery that exposed the worthlessness of Hamilton's testimony. It had to do with the spiral grooving, or rifling, inside a gun barrel. The raised parts between the grooves, known as lands, are what leave the distinctive and individual markings on bullets. Poser saw that Stielow's revolver displayed a conventional rifling pattern of grooves and lands, whereas the murder bullets had been fired from a gun that had an abnormal rifling pattern. Put in simple terms, the murder weapon had a manufacturer's flaw, and Stielow's did not. Presented with this incontrovertible evidence, Bond and Waite unhesitatingly recommended a pardon, and on May 9, 1918, Charles Stielow and Nelson Green walked out of Sing Sing as free men.

King and O'Connell were never tried for the West Shelby murders. After King had once again retracted his confession, the Orleans County grand jury, mindful of limited local enthusiasm and funds for another lengthy trial, declined to return indictments, and the matter was quietly forgotten.

Conclusion

What happened to Stielow and Green was grotesque but not entirely in vain. Their ordeal, and in particular the shoddy testimony that had occasioned it, so outraged Charles Waite that he devoted the rest of his life to improving the science of firearms identification. He founded the Bureau of Forensic Ballistics in New York, and saw his partner, Phillip O. Gravelle, design perhaps the single greatest advance in the field of bullet analysis—the comparison microscope, in which two halves of separate bullet images can be joined under the same lens, enabling the observer to closely compare the marks on each. Waite's dream of cataloging every gun in existence proved beyond him, but by the time of his death in 1926, his place in forensic history was assured.

Nicola Sacco and Bartolomeo Vanzetti

DATE: 1920

LOCATION: South Braintree, Massachusetts

SIGNIFICANCE: In their rush to bestow martyrdom on Sacco and Vanzetti, those who sought political capital from this tragedy conveniently ignored the overwhelming ballistics evidence.

On the afternoon of April 15, 1920, outside a shoe factory in South Braintree, Massachusetts, security guards Frederick Parmenter and Alessandro Berardelli were transferring the company's $15,777 payroll when two men approached. Without warning, one of the strangers opened fire, mortally wounding both guards. His partner, who wore a dark handlebar mustache, pumped yet more rounds into the helpless victims. Heaving the payroll boxes into a waiting car that contained three other men, the killers made their escape. Eyewitnesses described the gang as "Italian looking," but of more use to investigators were the empty shells recovered from the sidewalk. All were manufactured by three firms: Peters, Winchester, and Remington.

Two days later, a stolen Buick thought to be the getaway vehicle was found abandoned in some woods. Evidence linked it to an aborted payroll robbery at another shoe factory in nearby Bridgewater the previous Christmas Eve. It was believed to have been masterminded by an Italian named Mike Boda, but when police raided Boda's suspected hideout, he had already fled.

However, two other men were arrested: Nicola Sacco, twenty-nine, and his mustachioed companion, Bartolomeo Vanzetti, thirty-two. Both denied owning any guns, yet each possessed a loaded pistol, and Sacco's was a .32, the same caliber as the murder weapon. Also, Sacco was carrying twenty-three bullets, all made by Peters, Winchester, and Remington.

Vanzetti was a fish peddler; Sacco—significantly—worked in a shoe factory. Both were members of anarchist cells that openly espoused violence, a fact that inflamed public opinion against them. Identified as a participant in the bungled Bridgewater robbery, Vanzetti was jailed for ten to fifteen years. Sacco was not tried for the robbery, but he later joined his partner in facing charges of double murder.

Witch Hunt?

Eleven months later, on May 31, 1921, their trial opened in Dedham, Massachusetts, amid the hysteria of America's first "Red Scare," a time when anyone whose politics even hinted of radicalism was considered dangerously subversive. Judge Webster Thayer, who had presided over Vanzetti's earlier trial, again sat on the bench. Although much has been written about Thayer's alleged prejudice toward the accused, as the court records show, regardless of what he may have said or thought in private, his judicial conduct was never less than fair. In contrast, the defense team cast scruples to the wind. They had used the interim well, forging a coalition of anarchists, communists, and various labor unions, all under the umbrella of the Sacco-Vanzetti Defense Committee and all prepared to ruthlessly exploit the defendants for their own selfish ends. The trial was, they jeered, a witch hunt. This was palpably untrue, but the notion festered—particularly abroad—that Sacco and Vanzetti were being tried more for their politics than for any crime.

The court heard dozens of identification witnesses, fifty-nine for the prosecution and ninety-nine for the defense, a welter of testimony that produced only confusion. Similar ambiguity surrounded the question of whether Sacco's .32 had actually fired the bullet that killed Berardelli. Whereas one prosecution expert declared that it was indeed the murder weapon, another would only concede the possibility. Two defense experts, James Burns and Augustus Gill, harbored no such doubts, being adamant that Sacco's gun could not have fired the fatal bullet.

Any ambiguity raised by the gun paled in the face of one incontrovertible and damning fact: the bullet that killed Berardelli was so outdated that the prosecution's expert witnesses could locate none like it to test Sacco's gun—except the equally obsolete bullets from Sacco's pockets. On July 14, 1921, the jury returned a guilty verdict, and Judge Thayer sentenced the defendants to death.

The outcome touched off a firestorm of protest. Around the globe, left-wing parties lionized Sacco and Vanzetti, portraying them as innocent victims of capitalist justice. With the rhetoric on high boil, out of the shadows stepped Albert Hamilton, still smarting from the Stielow fiasco and desperate to restore his flagging reputation. Commissioned by defense lawyers to examine the disputed bullet, Hamilton announced with his usual pomposity that beyond any doubt it had not come from Sacco's gun.

With the clamor mounting, prosecution firearms expert Charles

Sacco and Vanzetti—canonized by the faithful, damned by the ballistics evidence. (National Archives)

Van Amburgh reexamined the bullets. By this time, in the fall of 1923, recent technological advances allowed Van Amburgh to enlarge photographs of the fatal bullet and bullets fired from Sacco's revolver. Again he concluded that they were identical.

During the motion for a retrial, events took a sensational turn when Hamilton brought two new Colts into court and disassembled them—along with Sacco's revolver—ostensibly to demonstrate some point. When the court's attention was diverted, he slipped one of the new barrels onto Sacco's revolver and tried to leave with the two Colts. Had it not been for Judge Thayer, who had followed the demonstration with lynx-eyed intensity, a grave miscarriage of justice would have occurred. Thayer caught Hamilton red-handed and ordered him to hand over the genuine barrel. Given such chicanery, it was hardly surprising that Thayer denied the motion for a new trial.

Persistent Critics

Still the outcry continued. In June 1927, a committee appointed to review the case contacted the man who would become America's leading firearms expert, Calvin Goddard, at the Bureau of Forensic Ballistics in New York. Armed with two recent inventions, the comparison microscope and the helixometer, Goddard traveled to Dedham. The helixometer, invented by physicist John H. Fisher, was a hollow probe fitted with a light and a magnifying glass for examining the insides of gun barrels. With defense expert Augustus Gill acting as witness, Goddard fired a bullet from Sacco's revolver into cotton wool, then placed it beside the murder bullet on the comparison microscope. The outcome was unequivocal—the murder bullet had been fired from Sacco's revolver. Gill, peering through the microscope, had to agree. He exclaimed, "Well, what do you know about that!" When his fellow defense expert, James Burns, also changed his opinion, Sacco and Vanzetti's last hopes were dashed. On August 23, 1927, over worldwide protest, they died in the electric chair.

Conclusion

Nevertheless, the controversy lived on. In October 1961, a forensics team led by Colonel Frank Jury, former head of the New Jersey Firearms Laboratory, reexamined Sacco's revolver and concluded beyond any doubt that it was the murder weapon. Yet another investigation, this one conducted in March 1983 and underwritten by a Boston TV station, again confirmed Goddard's findings, leading Marshall Robinson of the Connecticut State Police Ballistics Department to remark wearily, "Why do they keep running these tests over and over? They always come out the same." It was a question people had been asking for decades.

Frederick Browne and William Kennedy

DATE: 1927

LOCATION: Stapleford Abbots, England

SIGNIFICANCE: This case made all of Europe aware of ballistics evidence and established the reputation of Britain's trailblazing firearms expert, Robert Churchill.

In the years following World War I, America took a decisive lead in the fledgling science of firearms analysis. Europe's desultory progress was explained in large measure by the relative scarcity of guns on the street, but it was also due in part to a belief that armed gunmen were exclusively an American phenomenon, one most unlikely to cross the Atlantic. Such smug complacency was shattered by the events of September 26, 1927.

On that night, two petty crooks, Frederick Browne, forty-six, and William Kennedy, thirty-six, set out from London by train to Billericay in Essex. Their intention was to steal a particular car that Browne had earmarked earlier. Thwarted in that desire—a barking dog scared them off—they broke into a garage belonging to a Dr. Edward Lovell, stole his blue Morris Cowley, and sped erratically back to London.

Some miles along a remote country lane, their haphazard progress drew the attention of Police Constable George Gutteridge, who flagged the Morris to a halt. He approached the car, shone his lamp on both men, and asked where they were going. Nettled by Browne's superciliousness, Gutteridge reached for his notebook. As he did so, Browne drew a gun and fired twice. Gutteridge fell to the ground. Browne sprang from the car and stood over him. Perhaps mindful of the superstition that a murder victim's eyes record the last image they see, he leaned over the prostrate officer and shot out his eyes. Later that night, the two killers ditched the car in south London before catching a tram to Browne's Golden Globe Garage in Battersea.

First light saw a fast-moving chain of events: A motorist found Constable Gutteridge's bullet-riddled body lying beside the road, Dr. Lovell contacted the police to report the theft of his car, and the Morris was discovered in London. Investigators soon linked the three incidents.

The stolen car provided several promising clues. There were splashes of blood on the floor and running board, and under the passenger seat lay an empty cartridge case. This was handed to Robert Churchill for analysis. Churchill came from a family of London gunsmiths that made and sold sporting guns and rifles, but his real interest lay in the study of weapons and their projectiles. On those rare occasions when Scotland Yard needed to consult a firearms expert, his was the opinion in demand. He identified the cartridge as a Mark IV, an obsolete bullet filled with black powder and manufactured at the Woolwich Arsenal in 1914. On its base he noted a tiny raised imperfection, the result of a faulty breechblock on the gun that had fired it. Churchill said that the murder weapon was almost certainly a .455 Webley revolver.

Finding that revolver, though, proved tortuous. As one lead after another dried up, the investigation ground to a standstill. For three months, the deadlock continued; then came a vital clue. An ex-convict who was being questioned in connection with a string of car thefts angrily protested his innocence, claiming that the real culprits were actually two other crooks named Browne and Kennedy. Furthermore, he'd heard them brag about killing Constable Gutteridge.

Dual Arrests

There was plenty in Browne's record to suggest that he was capable of murder; he had a long history of violence and once, while imprisoned, had brutally attacked a guard. Detectives were understandably cautious as they staked out his garage. Their patience was rewarded on January 20, 1928, when their target drove up under cover of darkness. As Browne alighted from his car, they swooped in upon him. Inside the car's glove compartment was a revolver. More guns were found indoors, together with two thousand pounds (about eight thousand dollars) hidden in the lavatory cistern and medical instruments similar to those taken from the Morris Cowley.

Five days later in Liverpool, Kennedy was detained after a struggle in which he attempted to shoot the arresting officer (only the jamming of the gun saved the policeman's life). It was a strange reaction for someone who subsequently claimed that his role in the murder had been that of passive bystander. Browne, he said, had shot Gutteridge without any provocation. Bitterly contemptuous of his erstwhile partner's attempt to save his own neck, Browne dismissed the statement as a "concoction."

Within the arsenal found at Browne's garage was a .455 Webley revolver loaded with the same ancient ammunition that had killed Constable Gutteridge. When test-fired by Churchill, each cartridge revealed an identical breechblock imperfection. No fewer than thirteen hundred Webley revolvers were tested in efforts to replicate the flaw, but none ever did. Churchill also noted that the black powder loaded into the Mark IV ammunition was identical to powder traces tattooed into the skin around Gutteridge's wounds.

At their trial, Browne and Kennedy were apportioned equal guilt. On May 31, 1928, Browne was hanged at Pentonville Prison, while a few miles across London, Kennedy was similarly dealt with at Wandsworth.

Conclusion

For most, this case provided an unwelcome foretaste of an increasingly violent future, but mixed with it was an undeniable awe regarding Churchill's confidently delivered testimony. This trial made him a household name, a position he was to relish for the next two decades.

John Branion

DATE: 1967

LOCATION: Chicago, Illinois

SIGNIFICANCE: In this case, outstanding firearms analysis and meticulous detective work combined to crack a "perfect" alibi.

Just before lunch on December 22, 1967, Dr. John Branion, forty-one, set off in his car from the Ida Mae Scott Hospital on Chicago's south side. Several minutes later—after passing his home—he picked up his young son from nursery school, then called on Maxine Brown, who was scheduled to lunch with the Branions that day. When Mrs. Brown explained that she was unable to keep the engagement, Branion drove to his apartment at 5054 South Woodlawn Avenue. He arrived at 11:57 A.M. to find his wife, Donna, lying on the utility room floor. She had been shot repeatedly. He immediately summoned help. A neighbor, Dr. Helen Payne, examined the stricken woman and confirmed the obvious—Donna Branion was dead.

Police recovered three expended bullets and four cartridge casings. Two of the slugs were under the body and one near it. A fourth, still in the body, was found during the autopsy. Red dots on the shell casing primers were typical of German-made Geco ammunition.

A neighbor, Theresa Kentra, reported hearing muffled thuds at approximately 11:30 A.M. Another neighbor was even more precise. By chance, he had been making a long-distance telephone call and, mindful of the charges, had kept one eye on the clock. He, too, had heard what sounded like gunshots and instinctively noted the time—11:36 A.M.

Unusual Rifling

The bullets that killed Donna Branion were .38 caliber, quite common, but microscopic examination revealed distinctive rifling patterns—six lands and a right twist. The casings also had marks on the base, signifying that the weapon used to fire them had a loading indicator. Firearms expert Burt Nielsen knew of only one pistol that fulfilled all these criteria—a Walther PPK. When asked if he owned any weapons capable of firing .38 caliber ammunition, Branion, an avid gun collector, replied, "Just one," a Hi Standard. No mention was made of a Walther. Tests conducted on the Hi Standard eliminated it as the murder weapon.

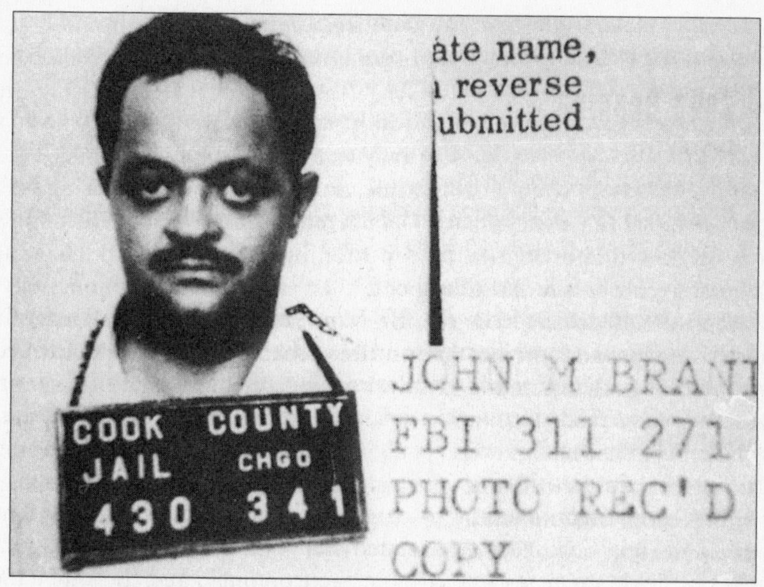

John Branion—murdered his wife in a hail of bullets. (FBI)

Detectives were puzzled by the lack of apparent motive for the killing. There was no sign of a robbery, and Donna was not known to have any enemies. Except possibly her husband. Rumors that the Branion marriage had been less than idyllic were commonplace. In a move that seemed to confirm the rumors, just forty-eight hours after the tragedy, Branion flew to Colorado for a Christmas ski vacation. In his absence, detectives learned that Branion was a notorious womanizer whose affairs had provoked numerous violent arguments with Donna. Compounding their suspicion was his behavior at the murder scene, where he had not bothered to examine his wife's body. He claimed that this was because he could tell from the lividity (the tendency of blood to sink to the lowest extremities in a corpse) that she was already dead. And yet, according to Dr. Payne, when she saw the body at 12:20 P.M., lividity was not present. Upon his return from Colorado, Branion hastily explained that he had really meant cyanosis, a blue discoloration of the skin caused by deoxygenated blood.

On January 22, 1968, Detective Michael Boyle returned to Branion's apartment with a search warrant. In a cabinet that had been locked on the day of the murder, he found a brochure for a Walther

PPK, an extra clip, and a manufacturer's target, all bearing the serial number 188274. He also found two boxes of Geco brand .38 caliber ammunition. One box was full; the other had four shells missing, the same number of shots that had killed Donna Branion.

The New York importers of the Walther revealed that model number 188274 was shipped to a Chicago store, where records showed that it had been purchased by James Hooks, a friend of Branion's. Hooks admitted giving the gun to Branion as a belated birthday gift almost a year before the killing.

Faced with this revelation, Branion abruptly changed tack. His original statement specifically mentioned that nothing had been stolen; now he blustered that the Walther must have been taken by the intruders who killed his wife. One change of heart too many led to Branion's indictment for murder.

Time Discrepancy

At his trial, prosecutors declared that the story Branion had told was correct in every respect save one—chronology. Yes, he had picked up his son and had then driven to Maxine Brown's. But first he had sneaked home and shot his wife. Joyce Kelly, a teacher at the nursery school, testified that Branion, normally so laid-back and casual, had rushed in, flustered and panting for breath, between 11:45 A.M. and 11:50 A.M., some ten minutes later than he had claimed.

But did Branion have sufficient time to commit the crime as hypothesized? Boyle described a series of tests that he and another officer had performed, driving the route allegedly taken by Branion. They had covered the 2.8-mile journey in a minimum of six minutes and a maximum of twelve—time enough, said the prosecution, for Branion to commit the murder, pick up his son, and establish an alibi. With that alibi in tatters, all the prosecution had to do was present the firearms evidence and Branion's fate was sealed. On May 28, 1968, Branion was convicted of murder and sentenced to a twenty- to thirty-year jail term.

Released on a cash bond of just five hundred dollars, Branion began an appeal process that lasted until 1971. With his legal options fast expiring and sensing that imprisonment was nigh, he fled the country. After an amazing jaunt across two continents, he found asylum in Uganda, occasionally acting as personal physician to Idi Amin, that country's dictator. Upon Amin's ouster, Branion was arrested and extradited to the United States in October 1983.

Conclusion

In August 1990, Branion was released from prison for health reasons. One month later, at sixty-four, he died of a brain tumor and heart ailment. To the end, he protested his innocence, but conspicuously absent from all his denials was any explanation for the firearms testimony.

Joseph Christopher

DATE: 1980

LOCATION: Buffalo and New York, New York

SIGNIFICANCE: Serial killers are usually creatures of habit; they find a method of destruction that works and stick to it. However, this extraordinary murderer employed no fewer than four different types of weapons to achieve his deadly aims.

In the fall of 1980, two separate but equally brutal murder sprees rocked New York State. The first began on September 22, when Glenn Dunn, a fourteen-year-old African American, was shot and killed outside a Buffalo supermarket. The next day, two more black males were gunned down. On September 24, a fourth was shot to death in nearby Niagara Falls. All were killed at close range by an "unidentified white youth" who carried a firearm in a paper bag. The bag served two purposes: besides concealing the weapon, it also caught the expended shell casings, a precaution suggesting that these were no mere random acts of violence but carefully planned slayings. However, in the final shooting, the gunman blundered. Unfamiliar with the Niagara Falls area, and in his haste to escape a crowd of pursuers, he dropped a shell casing. Cognizant of this slip—which was heavily reported in the Buffalo media—the ".22 Caliber Killer" switched his means of attack.

Two weeks later, on successive nights, two black cabdrivers were murdered and dumped in an isolated wooded area. One was killed by a screwdriver rammed through the back of his skull, a feat requiring unusual strength; the other was battered to death with a square-headed hammer. Both were stabbed several times through the sternum, and in a grisly addendum, each had had his heart cut out.

It is no easy matter to excise a heart, and yet the procedures had been performed deftly and without undue mutilation. This, combined with the assailant's rare bravado—very few serial killers will risk hand-to-hand combat with their victims—led investigators to suspect that they were seeking someone with hunting experience.

Whoever it was, he appeared to have suddenly lost his appetite for killing. Several weeks passed without incident. And then came an astonishing onslaught.

Bloodbath

Four hundred miles away in midtown Manhattan on December 22, within the space of a few hours, five blacks and one Hispanic were stabbed by a man who just ran up and attacked them on the street. Four died; the others sustained serious injuries. After the outburst, the "Midtown Slasher" vanished. At first, no one connected this murder spree with the earlier mayhem. Not until New York medical examiner Dr. Michael Baden compared photos of the Manhattan casualties—all slim and mustached—with those in Buffalo were the victims' physical similarities noted. As if to reinforce this belief, the Slasher reemerged in Buffalo, where another wanton knife binge left two more black males dead and several others wounded.

But already clues were surfacing. The paper bag had proved inadequate as a receptacle for the shell casings. Five were recovered from the first wave of killings, and these, together with the fatal bullets, were sent to the FBI Laboratory in Washington, D.C., for analysis. Experts identified five possible weapons that could have fired such ammunition—four handguns and a rifle.

In order to single out the murder weapon, technicians had to study the firing pin imprints. When a gun is discharged, its firing pin strikes the back of the bullet, leaving a distinctive and individual imprint on the casing. By obtaining illustrations of firing pin marks from all five gun makers, FBI experts eliminated three of the weapons immediately. The remaining two, a handgun and a rifle, were both manufactured by the Sturm Ruger gun company in Southport, Connecticut. Company specifications indicated that the murder weapon had to be a .22 rifle. Because no eyewitness had reported seeing the gunman carrying a rifle—the murder weapon had been hidden in a brown paper bag—this could only mean that the barrel had been sawed off.

GI Arrested

With the net tightening, a strange development occurred the following January. A recently enlisted soldier at Fort Benning, Georgia, was arrested on charges of slashing a black GI. After attempting to emasculate himself with a razor blade, the soldier, Joseph Christopher, twenty-five, was committed to a hospital. There he bragged to a nurse about killing numerous blacks in Buffalo and New York. A search of Christopher's belongings revealed a supply of .22 caliber ammunition, but more pertinent was the discovery that he had joined the army on November 13—which coincided with the lull in killing—and that he had been furloughed since December 19. A bus ticket logged his arrival in Manhattan one day before the killings began.

In the Buffalo house where Christopher lived with his mother and sisters, police searched the basement and found a single bullet that had misfired; the bullet and shell hadn't separated. On the casing was the firing pin imprint, an exact match with the cartridges found at the crime scenes. As anticipated, Christopher was an avid hunter, and at the family's hunting lodge forty miles southwest of Buffalo, police uncovered a virtual armory: boxes of .22 caliber ammunition, an extensive gun collection inherited from his father, and the sawed-off barrel of a .22 rifle. The only item missing in an otherwise overwhelming forensic case against the accused was the actual murder weapon. It was never found. In a videotaped confession, Christopher, not known to be a racist, revealed details of the killings that only the murderer could have known. The only motive he ever offered for his outbursts was that "something came over me. . . . It was something I had to do."

Sentenced to life imprisonment, Christopher was never tried for the murder of the two mutilated taxi drivers because prosecutors feared that putting such revelations before a jury might increase the chances of an insanity verdict and the concomitant risk of possible release at a later date.

Conclusion

By any standards, Joseph Christopher has to be considered one of the most remarkable killers of recent times, and his case underlines the degree of importance that modern courts attach to solid physical evidence. Two eyewitnesses to the killings, emphatic that Christopher was not the gunman, went largely unheeded, their testimony overwhelmed by the sheer weight of the forensic evidence that prosecutors had amassed.

James Mitchell

DATE: 1991

LOCATION: San Francisco, California

SIGNIFICANCE: Forensic acoustics and computer-generated animation have added yet another dimension to the ballistics expert's arsenal.

Over the course of two decades, brothers Artie and Jim Mitchell created a vast business empire out of the explosive California pornography market. By 1991, they were at the peak of their success, churning out low-budget, high-profit X-rated movies. They then plowed their vast profits into what became the most successful strip club in San Francisco. Both in their forties, they spent millions on a riotous lifestyle fueled by drugs and alcohol. Artie, in particular, lived at a reckless pace.

No one doubted their financial acumen—the figures spoke for themselves—but their management style was bizarre, to say the least. They traded punches and insults with each other with equal frequency. Muscle, not mediation, was the watchword when these two got together. Jim, two years his brother's senior and the more reserved of the two, adopted an almost paternal air toward Artie. But on a cold, raw San Franciscan night in February 1991, the two men had their final argument.

At 10:15 P.M. on the twenty-seventh, police were summoned to Artie Mitchell's house by his girlfriend, Julie Bajo. There had been a shooting, she screamed to the 911 operator. While the call was being taped, further shots could be heard in the background. Officers arriving at the house found Jim Mitchell wandering outside, apparently in a daze. He was carrying a .22 rifle and, in a shoulder holster, an unfired .38 Smith & Wesson Special. In the bedroom, Artie lay dead. He had been shot through the abdomen, right arm, and right eye. In all, police found eight spent .22 shell casings. That Jim fired the bullets was beyond dispute—he readily admitted as much—but was it premeditated murder rather than spontaneous manslaughter? The authorities clearly thought so and charged him with first-degree homicide.

Prosecutors decided to pin their case on the 911 tape; the pattern of recorded gunshots, they said, demonstrated a clear intent on Jim Mitchell's part to take his brother's life. To bolster this assertion, they called Dr. Harry Hollien, a Florida expert in forensic acoustics, to the stand. He explained how he had studied the tape, attempting to isolate and identify each shot. This involved replicating, as closely as possible, the room in which the shooting occurred. Curiously enough, Hollien found just such a room in his own house, a coincidence that the skeptical defense team found hard to accept.

Deadly Gap

Undeterred, Hollien produced spectrograms of the various test shots, then showed how he had compared them to the sounds on the tape. This allowed him to conclude that the first shot was fired 5.5 seconds into the tape; shot two, 10 seconds; shot three, 25.5 seconds; shot four, 54 seconds; shot five, 55.9 seconds. The half-minute gap between the third and fourth shots was central to the state's contention that Mitchell had waited for some time and then deliberately gunned down his already wounded brother. If so, it was a telling point. The major difference between murder and manslaughter is one of intent: if Hollien's interpretation of this tape was correct, then it undermined the defendant's claim that the shooting had happened in the heat of the moment.

On January 24, 1992, the prosecution submitted a computer-generated video animation of the shooting. It showed Artie approaching his bedroom door and walking down the hall, where he was shot twice, then entering the hallway and getting shot the final time in the head. The shots were represented by red laserlike lines against a blue background. Not only was this tape supposed to have reproduced the crime scene, it also purportedly showed the order in which the eight shots were fired. Clearly, this was moving into the realm of speculation. Only over strident defense objections, and after the tape had been modified to make the characters less lifelike, was the animation admitted into evidence.

Arizona criminalist Lucien Haag, the expert whose crime scene reconstruction had inspired the video, explained the methodology behind his conclusions. He had traced the bullet paths through the various rooms, noting angles and impact points. Because one bullet had passed through a door, he went out and purchased a door of the same type, just to measure what level of deflection it might exert. Under cross-examination, Haag admitted that the video was a subjective interpretation and was forced to concede that with so many shots there were thousands of possible sequences to consider.

This point was driven home by defense witness Charles Morton, another expert on crime scenes. Circumspect by nature, Morton felt that the whole subject of reconstructions was a forensic quagmire. He argued that the criminalist's role was to analyze the physical evidence available and to present any conclusions to the court in a fair and impartial manner. Thereafter, it was up to the jury.

On February 18, 1992, Jim Mitchell was acquitted of murder but found guilty of manslaughter. He received a six-year jail term.

Conclusion

The controversy surrounding computer-generated reconstructions is unlikely to diminish. As such representation gains acceptance, its objectivity will remain in dispute.

CAUSE OF DEATH

With the discovery of a dead body, three questions must be answered: Who is the person? How long has he or she been dead? What killed him or her? Often this last question has aroused some of the most celebrated confrontations since science first started taking an interest in crime in the eighteenth century. Clearly how a person died is of paramount importance in determining whether there is a criminal case to pursue and ultimately whether charges are filed. For instance, if a husband says that his wife incurred her lethal injuries falling down the stairs, and it could be shown that those injuries were inconsistent with such a fall, then that husband would have some explaining to do.

This need for absolute accuracy makes the initial autopsy critical. The primary onus for establishing cause of death lies with the pathologist, and in the overwhelming number of cases the cause of death is readily apparent, often gruesomely so. But if nothing organic indicates how a particular person died, then samples of body tissue and fluids must be taken for analysis. Although the standard of modern-day analysis has reached quite spectacular levels, the ever-increasing range of synthetic substances that inhabit our daily lives makes vigilance the watchword in matters of detecting exotic poisons. In this section, all of the cases generated lengthy and often acrimonious debate as to what exactly the victim died of. None was straightforward; all are absorbing.

Norman Thorne

DATE: 1924

LOCATION: Crowborough, England

SIGNIFICANCE: This is the first case in which the opinion of Britain's foremost pathologist, Sir Bernard Spilsbury, was called into question.

Ever since 1922, Norman Thorne had struggled to eke out an existence from a run-down poultry farm on the outskirts of Crowborough in Sussex. He not only raised chickens but lived among them as well, in a squalid shed barely fit for human habitation. At twenty-four, he was perfunctorily engaged to a young typist from London, Elsie Cameron, a plain and neurotic girl with but one thought in life—to find a husband. Some idea of Elsie's desperation can be gauged from her willingness to bed down with Thorne in his filthy shack, which she did in hopes that it would lead to something more permanent. But Thorne had other plans. Each morning after she had stayed with him, he would shunt Elsie back to London on the first available train. By letter she pressed him to announce a wedding date. In November 1924, she changed tack, writing untruthfully that she was pregnant. Thorne was unmoved, announcing that he intended to marry another girl, a local beauty named Elizabeth Coldicott. In a fury, Elsie wrote Thorne that she would be arriving on the afternoon of December 5, at which time she expected him to do the decent thing. Fiercely determined, and with all her worldly possessions crammed into a single bag, she caught the train from London.

Five days later, her anxious father telegraphed Thorne to find out what had become of her. Thorne expressed bewilderment, saying that Elsie had not arrived. When Mr. Cameron contacted the Crowborough police, they interviewed Thorne and came away impressed by his concern and obvious desire to help. Soon Elsie's disappearance attracted reporters from Fleet Street. Thorne was good copy, as he enlarged on the plans that he and Elsie had made to marry and agonized over her welfare. An accompanying photograph taken of him in the chicken run, mournfully scattering seed to his flock of leghorns, only emphasized his isolation.

Walked to Her Death

However, two men who knew Elsie told the police they had seen her walking toward Thorne's farm on the night of December 5. Strangely, not until a month later, when a vacationing neighbor returned and told the same story, did Scotland Yard become involved. Thorne repeated his earlier story to the Yard detectives on January 14, unaware that digging had already begun at his farm. The following day, Elsie's bag was discovered. After some consideration, Thorne made a statement. Elsie had arrived, he said, telling him that she intended to stay until he married her. An argument broke out, and Thorne had stormed from the shed. Upon his return, he found Elsie swinging by a noose from the hut's main beam. In a blind panic, Thorne opted to chop his lover into quarters and bury her in the chicken run, at the very spot where his photograph had been taken.

Elsie's remains were dug up and taken to the local mortuary for examination by Sir Bernard Spilsbury. Naturally, he devoted close attention to the neck. In suicidal hanging, death is caused by rupture or obstruction of blood vessels in the brain due to pressure on the vessels of the neck, obstruction of the windpipe, or both. Usually the rope or ligature used leaves a deep bruise on the flesh at the point of pressure. Spilsbury found none of these features, nothing to suggest that Elsie Cameron had hanged herself. What he did find were bruises on her head, face, elbow, legs, and feet "sufficient to account for death from shock." In other words, she had been bludgeoned to death. On January 26, 1925, the remains were reburied, and Norman Thorne found himself facing a charge of murder.

Realizing that their sole hope lay with the suicide story, Thorne's defense team demanded a second autopsy, this time to be carried out by Dr. Robert Bronte, then pathologist at Harrow Hospital and a former Crown Analyst to the Irish government. Bronte loathed Spilsbury, so it was with some tension that in late February they met at Willesden Cemetery to preside over the second disinterment of Elsie Cameron's body. Spilsbury studied Bronte's autopsy with gimlet-eyed intensity.

At the subsequent trial, everything hinged on one question—how did Elsie Cameron die? Spilsbury insisted she had been battered to death. Bronte, while conceding that his examination had suffered because of the advanced state of decomposition, pointed to marks on the neck as proof that death had resulted from suicide through hanging. Spilsbury dismissed these as "natural creases" in the neck. Not so, argued Bronte; they were clearly "grooves" and, furthermore, displayed signs of bruising. He had taken slides of the grooves and passed sections of the bruised areas to Spilsbury so that he might make his own slides. Spilsbury testified that neither his slides nor those of Bronte showed any signs of bruising.

Seriously undermining Bronte's argument was one critical fact. Along the main beam in Thorne's hut—the one he said that Elsie hanged herself from—was a thick layer of undisturbed dust. Had a rope been tied around that beam and then used to suspend a body, not only would it have dislodged the dust, but the beam itself would have shown traces of grooving. There were no such markings.

The judge, in summarizing, described Spilsbury's opinion as "undoubtedly the very best . . . that can be obtained," and the jury agreed. Thorne was sentenced to death, an outcome that aroused considerable public outrage. But the appeal was unsuccessful, and on April 22, 1925, the failed farmer turned bungling murderer went to the hangman's noose.

Conclusion

Over the years, Dr. Robert Bronte, irascible and garrulous, never missed an opportunity to castigate Spilsbury, often couching his criticisms in the most unprofessional of terms. Spilsbury, for his part, maintained the lofty dignity that so impressed judge and jury alike, all derived from the knowledge that not once in court was he ever bested by his excitable rival.

David Marshall

DATE: 1926

LOCATION: Philadelphia, Pennsylvania

SIGNIFICANCE: As science extended the parameters of detection, it also raised the temperature of courtroom disputes.

Dusk had just fallen on downtown Philadelphia when Anna May Dietrich bid farewell to her sister after a shopping expedition, saying she was on her way to a dancing lesson and would be home for dinner that night. It was an appointment that the thirty-five-year-old milliner never kept. The next morning—January 19, 1926—her family alerted the police. They contacted the dance academy and were told that Anna had canceled her appointment because she had to see a friend. Anna's brother-in-law, Alexander Schuhl, wondered if that friend might be David Marshall, whom the missing woman had known for some years.

Marshall, an unregistered chiropractor, was helpfulness itself when officers called. No, he had not seen Anna for more than a week, but he did remember her being greatly distressed, something about a soured romance. . . . Police thanked Marshall for his assistance and left.

The following day, a woman on her way to catch a trolley in Media, a suburb of Philadelphia, noticed scraps of bloodstained newspaper lying alongside the road. Her eyes were drawn to some undergrowth and what appeared to be a bundle. Rather than look more closely, she ran for assistance in the form of William Rowson, a local blacksmith. Rowson gingerly opened up the parcel. Inside, wrapped in two newspapers, was the headless and legless torso of a young woman in a blue serge dress. Close by was another parcel containing the severed legs. Later that day, Schuhl identified the dress as one worn by his missing sister-in-law. A lack of blood, either in the body or on the clothing, suggested that the body had been drained by someone with obvious surgical knowledge before being dumped. Two days later, a human head, also wrapped in newspaper, was found some miles away. The frozen features were easily recognizable as those of Anna May Dietrich.

At this point a woman came forward, claiming that she had seen Anna in tears at Marshall's office on the night of her disappearance. But it was what chauffeur E. J. Barry had to say that really ignited a fire under the investigation. He told the head of Philadelphia's homicide division, William Belshaw, that on the night of January 20, Marshall had hired him to haul away some parcels from his surgery. As he lifted one of the packages, its paper wrapping broke and out fell a human leg. Barry stood aghast. Marshall frantically began thrusting fistfuls of dollars at him, begging him to get rid of the parcel. But Barry would have none of it and left.

Such a revelation convinced Belshaw that a second chat with the amiable chiropractor was in order. This time he escorted Marshall to the morgue and confronted him with the grisly remains. District Attorney William Taylor first made everyone bare their heads, then said to Marshall, "In the presence of God and this girl's body, didn't you do this?" Marshall stroked his mustache, lit a cigar, and smiled. "Why, certainly not."

Third Degree

After several hours of what newspapers referred to as a "severe grilling," Marshall said that Anna had committed suicide, using poison she had found in his surgery. He had cut up her body in order to dispose of it. Another night of questioning brought forth the full facts. Marshall confessed all. He told of having had a seven-year-long affair with Anna. He said that she had come to his surgery on January 19 determined to expose their relationship to his wife, demanding money and screaming, "You ruined me; I'll ruin you!" To stifle her yelling, he had stuffed a handkerchief into her mouth, at which point she had become faint and then collapsed, dead. At no time, he said, had he intended to kill her, which made his next course of action seem rather bizarre, to say the least—he cut her throat.

That night he left the body at his surgery and went home, but insomnia got the better of him. He returned at 8:30 A.M. and commenced reducing his former mistress to easily transportable pieces. The ghastly chore lasted until midday and was interrupted only once, by a phone call from Mr. Schuhl, anxious to learn his sister-in-law's whereabouts. That afternoon, Marshall drove around Delaware County, scattering segments of the body as he went. The next day, he disposed of the head beside some railroad tracks, where it was found by an inquisitive dog being exercised by its owner. To allay the memory of his grim labors, Marshall treated his wife to a night at the theater.

At his trial, which opened on March 8, 1926, Marshall recanted the confession, claiming it had been beaten out of him by overzealous investigators. He reiterated his claim that Anna had swallowed poison and that he had panicked and cut up the body to conceal his illicit relationship with her. It soon became apparent that this case would hinge on whichever side managed to win the forensic battle. Prosecution witness Dr. Clarke Stull, who conducted the autopsy, would not be deflected from his belief that Anna May Dietrich had been strangled, a view shared by Dr. Henry Wadsworth. For the defense, Dr. Henry Cattell maintained that nothing in the autopsy results was inconsistent with Marshall's version of events. Had Miss Dietrich been strangled, Cattell said, then marks would have been left on her neck; he could find none. Neither did he rule out Marshall's claim that the victim had taken poison. In rebuttal, the state commissioned Dr. J. Atlee Dean, chemist and bacteriologist, to examine the dead woman's organs. He testified that there was nothing to suggest poisoning.

During these exchanges, the defendant was often reduced to the status of a mere bystander. But steadily the prosecution made headway. When the case went to the jury on March 24, the result never seemed in doubt. Some five hours later, as an indication to the tipstaff that their labors were complete, the jury burst into an impromptu chorus of "Show Me the Way to Go Home." They convicted Marshall of second-degree murder.

Ten years later Marshall was paroled and moved to Florida, where he died soon after.

Conclusion

In all probability, courtroom disapproval of Anna Dietrich's unconventional social life kept Marshall out of the electric chair; indeed, had he not dismembered her body, in all probability he would have been tried on charges of manslaughter. While the glib chiropractor's adultery was glossed over with barely a mention, Anna Dietrich's morals drew condemnation from every quarter.

James Camb

DATE: 1947

LOCATION: West Africa

SIGNIFICANCE: Had the victim died from a fit or had she been strangled? Ordinarily such facts are not in dispute, but this was far from an ordinary case.

Eight days after setting out from Cape Town on October 10, 1947, the ocean liner SS *Durban Castle* was steaming steadily northward through the tropical night, some ninety miles off the West African coast, when galley steward Frederick Steer was summoned by a bell from cabin 126 on B deck. The time was just a few minutes before 3 A.M. Arriving at the cabin, he noticed that both the steward and stewardess lights above the door were illuminated, which struck him as odd, because passengers normally called one or the other. When Steer knocked, the door was opened a few inches and then slammed shut in his face by a half-dressed man who mumbled, "It's all right." Although the incident lasted only a split second, Steer instantly recognized the man as fellow crew member James Camb, a thirty-one-year-old deck steward and something of a nautical gigolo.

Still puzzled, Steer contacted the senior steward on duty, James

Murray, and together they returned to the cabin. They listened outside for a few minutes, but all seemed quiet, so they left. Murray thought it just as well to report the incident to the bridge, but without mentioning that Camb had been in the cabin because such a flagrant breach of regulations could result in the steward's dismissal. The bridge officer, well used to the nocturnal shenanigans of liner passengers, decided to take no action.

Just a few hours later, at 7:30 A.M., the stewardess who regularly cleaned cabin 126 was surprised to find the door open. Normally the occupant, Gay Gibson, an attractive twenty-one-year-old actress returning to London after a brief stint on the South African stage, kept the door locked. Hesitantly, the stewardess knocked and entered. Miss Gibson was nowhere to be seen. Her bed looked unusually disheveled, and the stewardess noticed stains on the sheet and pillowcase. As time wore on, with no sign of Miss Gibson, the stewardess reported her absence to the bridge. After a thorough but fruitless search of the ship, the captain ordered that the *Durban Castle* be put about in case the missing passenger had fallen overboard. In such a vast expanse of ocean, hopes of recovering anyone alive from the shark-infested waters soon faded, and the ship resumed its northerly route.

As rumors surrounding the disappearance rocked the ship from stem to stern, inevitably the name of James Camb arose. Summoned to the bridge and asked to account for his actions the previous night, he admitted talking to Gay Gibson on deck but vehemently denied entering her cabin. At the ship surgeon's suggestion, he submitted to a physical examination that revealed scratches on his shoulders and wrists, the result, Camb claimed, of heat rash.

Because news of Miss Gibson's disappearance had been radioed ahead to England, when the *Durban Castle* docked in Southampton, Camb was held for questioning. At first he persisted in his story, but then he admitted that he had been in the missing woman's cabin. The assignation had been arranged, he said, and Miss Gibson had greeted him wearing only her dressing gown. Underneath she was naked. During intercourse, Camb claimed she had suddenly started gasping for breath, then fell limp. "Her mouth was a little open . . . there was a faint line of bubbles . . . just on the edges of the lips. It . . . appeared to be blood-flecked."

Pushed Body Through Porthole

According to Camb, he had attempted artificial respiration, but when this failed he had lost his head and pushed the dead woman through the porthole. He could not explain how the bell pushes in the cabin had been pressed. The officers interviewing Camb thought they knew. Far from being a willing partner, Gay Gibson had tried to fight off the sex-crazed steward and had sought to summon help by calling both stewards. Overcome with lust, Camb had either strangled or smothered Miss Gibson, then got rid of the evidence, believing that without a body it would be impossible to prove murder.

If this was Camb's thinking, he was grievously mistaken. When his trial opened on March 18, 1948, the charge read "murder," and the prosecution felt confident that they could prove their case beyond a reasonable doubt. The circumstantial evidence weighed heavily against the accused. If, as he claimed, Gay Gibson had been naked beneath her dressing gown, then why were her pajamas missing from the cabin? And had she been expecting to engage in sexual activity, then surely she would have used the diaphragm contraceptive that was found in her suitcase. Then there were the two bedsheets from cabin 126. Both showed clear traces of blood group O, as well as smears of lipstick and

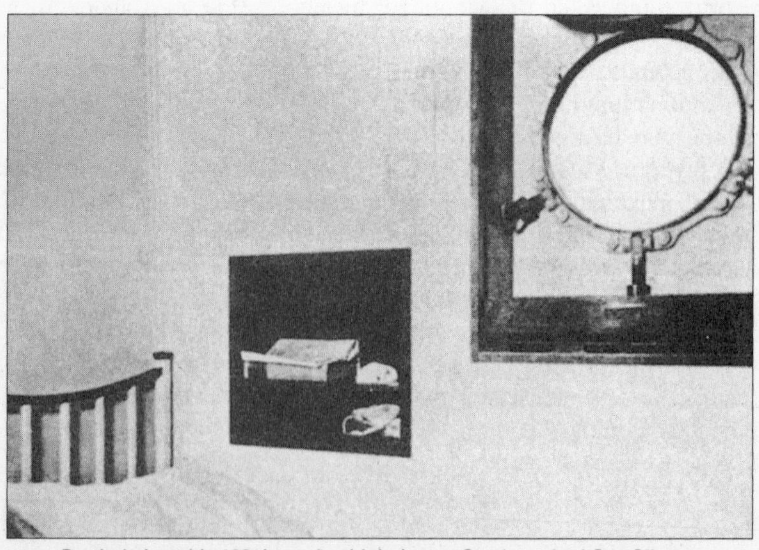

Porthole in cabin 126 through which James Camb pushed Gay Gibson.

saliva. Because Camb's blood group was A, it was reasonable to assume that it had come from the missing woman. Pathologist Dr. Donald Teare testified that while the bloodstains were consistent with manual strangulation, he would have expected to find traces of urine on the sheet, because in cases of strangulation the bladder commonly discharges its contents. On this occasion Dr. Teare had been ill served by his analysts. There was urine on the sheet, but it took a defense witness to find it!

Dr. Frederick Hocking and Professor James Webster both opined that Camb's version of events was not wholly impossible, but with impeccable fairness Hocking told how his tests had discovered the presence of urine on the top sheet; and within this sample he had also isolated human cells of the type that line the external female sex organs. Had the events unfolded in accordance with Camb's story, he would have been covered in urine as the dying woman thrashed beneath him. Yet he made no mention of this. All of which tended to substantiate the prosecution's view that Gay Gibson had been strangled and dumped overboard because she rejected Camb's sexual advances.

This was clearly the version the jury believed, and after only forty-five minutes of deliberation, they found Camb guilty of murder. Sentenced to death, he was later reprieved because capital punishment had been temporarily suspended while its future was being debated.

Conclusion

In becoming the first British defendant to be convicted of murder without a body being found, Camb made legal history. Only after the trial was it revealed that he had a history of attacking female passengers, though none had ever pressed charges. Paroled in 1959, he was convicted eight years later of molesting a thirteen-year-old girl. Inexplicably, he received only probation. Two years later, he was again convicted of sexual offenses against schoolgirls and returned to prison to serve out the remainder of his life sentence. He died on July 7, 1979.

Kenneth Barlow

DATE: 1957

LOCATION: Bradford, England

SIGNIFICANCE: In this landmark case, a panel of doctors, chemists, and forensic experts sought to prove that willful murder had been done.

At around midnight on May 3, 1957, a doctor was summoned to the Yorkshire home of thirty-eight-year-old Kenneth Barlow. When the doctor arrived, Barlow had a tragic tale to tell. All night long, his wife, Elizabeth, had been ill. At 9:20 P.M., while in bed, she had vomited. Barlow had changed the sheets, then joined his wife in bed. Sometime later, she had complained of "feeling too warm" and got up to take a bath. Barlow dozed off to sleep. When he awoke at 11 P.M., he found that Elizabeth was not beside him and hurried to the bathroom. There he had found her submerged in the water. At first he had tried to pull her out, but she was too heavy for him. So he had removed the plug and tried to revive her with artificial respiration.

Elizabeth still lay in the empty bath on her right side. Although there were no signs of violence on the body, the pupils were strangely dilated, a feature that the doctor thought worthy of investigation. For this reason, he contacted the police. Barlow smoothly repeated his story to them. They listened with interest, all the while wondering how someone who claimed to have made "frantic efforts" to haul his wife from the bath had managed to keep his pajamas so dry and avoid splashing the bathroom floor. Another incongruity was spotted by Dr. David Price, the medical examiner—Elizabeth Barlow still had water in the crooks of her elbows, hardly likely if she had received artificial respiration.

Two hypodermic syringes were found in the kitchen, which Barlow, a nurse, explained by saying that he had been giving himself injections of penicillin to treat a carbuncle. He denied giving his wife any injections. Traces of penicillin in the needles seemed to bear out Barlow's story.

An autopsy revealed that Elizabeth Barlow had been a normal, healthy woman, and there were no visible injection marks on the skin. She was two months pregnant, but Price could find nothing to account for the sudden onset of fainting in the bath. Analysis of the bodily organs told much the same tale: no trace of poison or any other metabolic weakness likely to result in loss of consciousness.

Murder Marks

On May 8, still dissatisfied, Price went over every inch of the dead woman's skin with a magnifying glass, looking for injection marks of a hypodermic needle. Mrs. Barlow's freckly complexion made this an arduous task, but after two hours of painstaking inspection, Price found two tiny puncture marks on the left buttock and another two in a fold of skin under the right buttock. Cutting into the skin and tissue around the marks, Price saw the minute inflammation consistent with recent injections.

But what substance had been injected? A council of doctors and scientists from around the country, headed by Dr. Alan S. Curry of the Home Office Forensic Science Service, was convened to consider the baffling facts. Barlow, the nurse, had efficiently described his wife's symptoms—vomiting, sweating, weakness, and pupil dilation. After much debate the panel agreed that everything pointed to hypoglycemia, or low blood sugar, a disorder that can, in extreme cases, lead to death. Had Barlow injected his wife with a massive overdose of insulin, then her blood sugar could have plummeted to a lethal level. All of which sounded plausible, except that Elizabeth Barlow's heart blood had registered a sugar level way above average—completely opposite what might have been expected.

Despite this setback, the panel would not be dissuaded from its belief that insulin—for the first time—had been used as a murder agent. They knew that Barlow frequently injected insulin at work and that he had once joked to a patient: "If anybody ever gets a real dose of this, he's on his way to the next world." Another comment was even more enlightening. Barlow had confidentially advised a fellow nurse that insulin was the ideal choice for a "perfect murder," because it dissolved in the blood and could not be traced.

He was right—there were no prescribed tests for detecting insulin in the body—but eventually the panel solved the conundrum of Elizabeth Barlow's high sugar level. In several cases of violent death, biochemical research had shown that the liver often floods the bloodstream with sugar in the last few moments before death as a survival aid. If this reaches the heart before circulation stops, then the blood there registers an unusually high blood sugar level. Which meant that Mrs. Barlow could have been given an insulin overdose.

Historic Experiment

To confirm their hypothesis, the team conducted an unusual experiment. First, a number of mice were injected with insulin. They trembled, made feeble noises, became comatose, and died. Then other mice were injected with extracts of the tissue surrounding the injection marks on Mrs. Barlow's body. Exactly the same reactions were observed. It was noted that mice injected with matter from the left buttock died more rapidly than those given tissue from the right, suggesting that the left injection had been administered last. The data gathered from this experiment confirmed that the quantity of the insulin remaining in the body was eighty-four units, although the actual dosage must have been much higher.

But what of the commonly held belief among doctors—and Barlow—that insulin disappeared very quickly from the body? Once again, new research came to the aid of the examiners. It was known that acidic conditions preserve insulin, and it now appeared that formation of lactic acid in Mrs. Barlow's muscles after death had prevented its breakdown.

Bradford police had already discovered that Barlow was no stranger to sudden death: just a year earlier, his first wife had died in mysterious circumstances at the age of thirty-three. The cause of death was never satisfactorily explained, and he had married Elizabeth soon afterward.

On July 29, 1957, Barlow was arrested and charged with murder. At first he persisted in denying that he had injected his wife at any time, until presented with the evidence. Then he admitted injecting her with ergonovine to induce an abortion. In fact, the forensic experts had already anticipated that very defense—no abortifacient drugs were found.

At Barlow's trial, the defense suggested that as Mrs. Barlow fainted and slid under the bathwater, her body had reacted by releasing a massive dose of insulin into the bloodstream, causing coma and death. This theory was briskly dealt with by Dr. Price. He reckoned that to account for the eighty-four units of insulin found in Mrs. Barlow's body, her pancreas would have had to secrete an incredible—and unheard of—fifteen thousand units!

Barlow was found guilty and imprisoned for life.

Conclusion

In his summing up, the judge paid particular tribute to the forensic team. It was a well-deserved commendation. Their employment of the section analysis had added another facet to modern forensic science.

Carl Coppolino

DATE: 1965

LOCATION: Longboat Key, Florida

SIGNIFICANCE: Prisons are full of people who committed the "perfect murder," as this particular killer found to his cost.

Just after dawn on August 28, 1965, Dr. Juliette Karow was roused from her sleep by a phone call. She listened as one of her patients, Carl Coppolino, himself a doctor, explained that he had just found his wife, Carmela, dead, ostensibly the victim of a heart attack. Karow frowned—young women in their thirties rarely suffer coronary failure, and Carmela had always seemed in perfect health—but she had no reason to disbelieve Coppolino, especially when he said that Carmela had been complaining of chest pains the night before. Karow drove the short distance to Coppolino's home on ritzy Longboat Key, just across the bay from Sarasota, Florida. Not long before, the Coppolinos had moved from New Jersey, where Carl, thirty-four, had been an anesthesiologist until ill health forced a premature retirement. Carmela, too, had been a doctor.

When Dr. Karow arrived, she found Carmela Coppolino beyond medical assistance. But she was deeply troubled. The position of the body seemed unnatural; Carmela lay on her right side, with her right arm tucked beneath her. Karow would have expected to find the hand swollen. It was not. Also, lividity did not seem consistent. Neither was the bedding rumpled. On the contrary, it seemed remarkably neat, staged almost. Despite her misgivings, Dr. Karow signed a death certificate, citing "coronary occlusion," then passed her findings to the Sarasota County Medical Examiner, who decided that there were no grounds for an autopsy. Just forty-one days later, Coppolino married wealthy socialite Mary Gibson, whom he had met at a local bridge club.

This development did not please fellow Longboat Key resident Marjorie Farber, a glamorous fifty-year-old widow who had migrated to Florida to be near Coppolino. Finally, after weeks of indecision, Mrs. Farber approached the only person she felt she could trust, Dr. Karow. She wanted to unburden her soul, she said. The story she told would keep newspapers across the nation in headlines for months to come.

It began when she and her husband, retired army colonel William Farber, had befriended the Coppolinos in New Jersey. In time, her association with Carl passed well beyond the usual bounds of good neighborliness, until by the evening of July 30, 1963, they were in the midst of a full-blown affair. That was the night William Farber was found unconscious in bed. Panic-stricken, Marjorie had pleaded on the phone for Carl to come over right away. Coppolino, who was drawing disability benefits for a supposed heart condition and therefore was not allowed to practice, sent Carmela over alone with the information that Farber had suffered convulsions the day before, the kind that often precede a heart attack.

First Body

Carmela found William Farber dead in the bedroom. Apart from being "all blue down one side," there was no outward sign of distress to the body. At Coppolino's urging, she dutifully signed the death certificate, listing "coronary thrombosis" as the cause.

That was the official version. Now, in Dr. Karow's surgery, Marjorie Farber insisted that every word had been a lie. The truth, she swore, was that Coppolino had given her a syringe filled with some solution and instructions to inject Farber when he was asleep. At the last moment her nerve had failed, but not before she had injected a minute amount of the fluid into Farber's leg. She summoned Coppolino to the house, and he finished off the job by strangling him. Dr. Karow immediately passed this story on to the proper authorities. Ordinarily, the claims of spurned lovers receive a skeptical hearing, but when it was learned that shortly before her death, Coppolino had increased the life insurance on Carmela to sixty-five thousand dollars, a discreet inquiry was launched.

Authorities in New Jersey and Florida obtained exhumation orders for both William Farber and Carmela Coppolino. The autopsies were carried out by Dr. Milton Helpern, New York's celebrated chief medical examiner. He examined the remains of William Farber and found no sign of heart disease but clear evidence of strangulation—the cricoid cartilage in the neck was fractured in two places.

Next, Helpern examined the body of Carmela Coppolino. Once again, he ruled out any coronary disease; Carmela's heart was in fine shape. Unfortunately for Helpern, so was everything else. Even an almost invisible hypodermic puncture mark on Carmela's left buttock didn't help. A battery of forensic tests revealed nothing. Helpern, veteran of more than twenty thousand autopsies, was baffled—until he considered Coppolino's former profession and asked himself this question: What drug would an anesthesiologist have access to that might cause untraceable death? The likeliest agent, he concluded, was an artificial form of curare called succinylcholine chloride.

Succinylcholine chloride causes complete muscular paralysis but does not induce unconsciousness, which meant that as Carmela's lungs refused to function, she would have been fully aware that she was suffocating but totally incapable of doing anything about it. Every textbook said that the drug was undetectable. Experts knew that it degraded to other chemicals in the body, but what those chemicals were no one had yet been able to establish.

Dr. Charles Umberger, chief of the medical examiner's toxicological department, listened to Helpern's problem and promised to work on it. In order to replicate as closely as possible the condition of Carmela's body at the time of autopsy, Umberger injected rabbits and frogs with a solution of succinylcholine chloride, then buried the carcasses and waited to see what would happen. Six months of patient experimentation enabled him to positively identify the chemicals that succinylcholine chloride degrades to in the body and their quantities.

He found an excessive amount of succinic acid in Carmela's brain—definite proof that she had received an intravenous injection of succinylcholine chloride sometime before her death. Significantly, just before Carmela's death, Coppolino had obtained considerable amounts of succinylcholine chloride from a colleague, explaining that he wished to conduct some experiments on cats.

Coppolino was at his home on Longboat Key when the police showed up to arrest him. Later, New Jersey announced that Coppolino would also be charged in the death of William Farber.

A Woman Scorned

Authorities decided to try him first in New Jersey, and after much legal fencing, Coppolino faced his accusers on December 5, 1966, in the Monmouth County courthouse. When Marjorie Farber testified, Coppolino's attorney, F. Lee Bailey, barreled in. There had been no murder, he roared; the entire episode had been a figment of this woman's malicious imagination, initiated by an evil desire for revenge on the man who had ditched her.

Bailey's attacks on Dr. Milton Helpern were more reasoned, less forthright. The main points of contention were whether William Farber had suffered from heart disease and whether the cricoid fracture had occurred before or after death. Helpern was emphatic on both counts, although Bailey drew from him the grudging admission that there was no bruising about the neck, as would normally have been present if strangulation had taken place.

Bailey attributed the cricoid fracture to rough handling of the body during disinterment, in particular a clumsy grave digger's shovel. Helpern dismissed such an idea. But Bailey had his own expert witnesses, and they thought otherwise. Doctors Joseph Spelman and Richard Ford, both experienced medical examiners, expressed the view that not only was the cricoid fracture caused after death but that Farber's heart showed clear signs of advanced coronary disease, certainly enough to have killed him.

Trial judge Elvin R. Simmill, in his final statement, commented on the vast array of conflicting medical evidence and stressed to the jury that they must be satisfied of Coppolino's guilt "beyond a reasonable doubt." It was an admonition that the jury took to heart. After deliberating for less than five hours, they returned a verdict of not guilty.

Coppolino's second trial—for the murder of his wife—began in early April 1967, and it was soon clear that the Florida prosecutors intended to make a much stronger case on the question of motive. State attorney Frank Schaub depicted the defendant as a money-grabbing philanderer, hell-bent on marrying Mary Gibson for her sizable fortune. Carmela's refusal to grant him a divorce had blown that idea sky-high. Instead, Coppolino began eyeing the insurance money, all sixty-five thousand dollars. With that, and Mary's bank account, he would be set for life.

This time, Marjorie Farber's testimony was of limited significance; everything hinged on the medical evidence. The jury listened closely as Helpern and Umberger described their unique experiments that had led to the discovery of poison in Carmela's body. It was a compelling, unanswerable performance. Significantly, unlike in the first trial, Coppolino opted not to testify on his own behalf.

Convicted of second-degree murder, Coppolino served twelve years of his life sentence before gaining parole in 1979.

Conclusion

In choosing succinylcholine chloride as an instrument of murder, Coppolino knew that his chances of detection were virtually nil. But he didn't take into account that science is always seeking—and finding—new ways to eradicate its own ignorance.

Richard Kuklinski

DATE: 1983

LOCATION: Orangeburg, New York

SIGNIFICANCE: Experts were sure that Kuklinski had poisoned several business partners. But how to prove it?

In September 1983 a man's body was found, trussed with tape and wrapped in plastic bags, in a wooded area of Rockland County, New York, just three miles north of the New Jersey border. There was a single bullet wound to the head. First indications were that the man had met his death recently, but when county medical examiner Dr. Frederick Zugibe began the autopsy, he noticed two peculiarities: the organs were fresh, and decomposition had started from the outside, which is the reverse of the normal process. Checking the heart, he discovered ice crystals, which supported his immediate suspicion that the body had been frozen, probably by a killer whose intention was to disguise the true time of death. (Had the body been thawed out before dumping, this might never have been established.)

Soaking the hands in water and glycerin made it possible to rehydrate the fingers and take prints. They were identified as belonging to Louis Masgay, a fifty-year-old Pennsylvania store owner who had disappeared on July 1, 1981. When found, he was still in the same clothes he had been wearing on the day he was last seen. This indicated that Masgay had been murdered that same day, and then his body had been kept, literally, on ice for two years before being dumped. Such callousness horrified investigators; clearly this was no ordinary killer.

They found out that on the day of his disappearance, Masgay had arranged to meet a New Jersey businessman to buy a batch of blank videos. Masgay, a cautious man, had hidden the ninety-five thousand dollars needed to complete the transaction behind a secret door panel in his Ford van. The van was later found abandoned, the secret panel ripped out, the money gone. Through phone records, investigators learned the name of Masgay's erstwhile business partner: Richard Kuklinski.

Kuklinski, a bearded, hulking bear of a man in his late forties, liked to portray himself as a currency speculator, but his true stock-in-trade was something altogether more sinister. Beneath a thin veneer of sophisticated respectability lurked one of the worst killers America has ever produced. He murdered for the mob, for himself, and always for money.

One Hundred Murders

Nobody knows how many people died at Kuklinski's hands, except Kuklinski himself. Some put the figure at more than a hundred. Although stories abound of him killing as early as high school, his first recorded victim was George Mallibrand, a three-hundred-pound wheeler-dealer from Pennsylvania. On February 1, 1980, Mallibrand made the mistake of arguing with Kuklinski over debts totaling fifty thousand dollars. Four days later, his bullet-riddled corpse was found stuffed into a fifty-five-gallon drum in Jersey City.

Anyone who dealt with Kuklinski was dicing with death. In 1982, he tempted Paul Hoffman, a New Jersey pharmacist, with an offer of some cut-rate hijacked ulcer medication. A shrewd judge of character, Kuklinski kept Hoffman dangling until the latter was practically begging him to finalize the deal. When the two men finally met on April 29, 1982, Hoffman was carrying twenty-five thousand dollars in cash—Kuklinski was carrying a gun. Neither Hoffman nor the money has been seen since. Kuklinski has hinted that the pharmacist also wound up in a concrete-filled oil drum.

Although by instinct a loner, Kuklinski occasionally teamed up with other thugs to run auto theft scams. In partners Daniel Deppner and Gary Smith, he chose badly. Neither was especially bright and both soon attracted the kind of police attention that Kuklinski went to great lengths to avoid. His reaction was typical. After sharing his concerns with the gullible Deppner, they agreed that Smith had to go. In a Bergen County motel where he had been holed up, Smith hungrily wolfed down the burger that Kuklinski had brought him. After a few mouthfuls the room began to spin. Kuklinski and Deppner roared with laughter as Smith choked on the cyanide burger. In the end, Kuklinski tired of waiting and strangled the hapless Smith, stuffing his body under the bed, where it was found four days later on December 27, 1982. During that time the room had been rented each night. Guests had wrinkled their noses at the smell, but none thought to look under the bed.

The following May, a bicyclist riding along Clinton Road in West Milford, New Jersey, noticed a huge turkey buzzard perched high in a tree. He went closer and saw a plastic garbage bag with a human head sticking out of it. Dr. Geetha Natarajan, the New Jersey medical examiner, performed the autopsy. A ligature mark around the neck revealed the apparent cause of death, although Dr. Natarajan noticed a pinkish lividity around the shoulder and chest that might have been caused by carbon monoxide poisoning. In the pocket of the man's jeans Dr. Natarajan found a wallet containing motel receipts and family photographs. Through these photos investigators identified the body as that of Kuklinski's other partner, Daniel Deppner.

As the body count mounted, the authorities were certain that Richard Kuklinski was a one-man killing machine, but they did not have a single scrap of evidence that would stand up in court. To remedy this deficiency, beginning in September 1986 an agent of the Bureau of Alcohol, Tobacco, and Firearms, Dominick Polifrone, went undercover and contacted Kuklinski in a sting that became known as Operation Iceman. After some initial sparring, Polifrone convinced Kuklinski that he could provide him with ten kilos of cocaine at $31,500 per kilo. At the same time he solicited suggestions on how to get rid of a "rich kid" who was proving troublesome. Unaware that every word was being taped, Kuklinski waxed lyrical on the merits of cyanide. "It's quiet, it's not messy, it's not noisy . . . there are even spray mists around . . . you spray it in somebody's face and they go to sleep . . ." He even described a test murder he had committed on the street, just walking along in a crowd with a handkerchief over his nose and spraying a man. The man collapsed and died, and everyone thought he'd had a heart attack. "The best way is to hit 'im right in the nose with a spray so he inhales it," Kuklinski said. "Once he inhales it, he's gone."

The Sting

Because Kuklinski was unable to get his hands on any cyanide at the moment—his supply had dried up—Polifrone agreed to provide the deadly poison. On December 17, 1986, the two met at a truck stop off the New Jersey Turnpike. As arranged, Polifrone brought with him three scrambled egg sandwiches and a jar of what was supposedly cyanide—actually quinine—to put on them. The plan was to meet the so-called rich kid at a motel and poison him.

Kuklinski took the sandwiches away to prepare them. The watching agents had not anticipated this, and, fearful for Polifrone's safety, decided to make the arrest early. Just minutes later, Kuklinski was taken into custody. Sure enough, the toxicology department at the New Jersey crime laboratory found "cyanide" (quinine) applied to the sandwiches. Kuklinski was charged with five murders.

His trial began on January 25, 1988. While the prosecution had a powerful case, it was entirely circumstantial. Defense claims that

Kuklinski had been merely bragging were supported by the fact that neither of the supposed poisoning victims, Smith and Deppner, showed any traces of cyanide during autopsy.

To counter this argument the prosecution put New York medical examiner Dr. Michael Baden on the stand. He explained how cyanide quickly degrades to carbon and nitrogen in the body, until after a few days you can't find it at all, which is why it's such a good murder weapon. Also, the distinctive smell of bitter almonds is discernible only on a fresh body; it vanishes with decomposition. But the cyanide didn't become entirely invisible—it showed up in the lividity. Fully conversant with the case history and autopsy photographs, Baden pointed out that the patches of red lividity present on both Smith and Deppner were consistent with cyanide poisoning.

Baden's testimony stressed only that the lividity could have been caused by cyanide; without the toxicology no one could say for certain. Even so, Kuklinski was convicted of murdering Deppner and Smith, and on May 25, 1988, he was sentenced to life imprisonment. In return for immunity for the rest of his family, he also admitted killing Mallibrand, Masgay, and Hoffman.

Conclusion

The assassin called The Iceman employed poison, guns, ligatures, crossbows, iron bars—anything that came to hand—in his extraordinary one-man killing business. After he disappeared into Trenton State Prison, most assumed that they had heard the last of this psychopath. But Richard Kuklinski was as hungry for publicity as he had been for money. And he was great copy. Twice he appeared in TV documentaries, and his alleged activities regularly lit up newspaper columns around the globe. Every appearance on either screen or page was calculated to promote the image of Kuklinski as the ultimate killing machine. As the years passed, his claims became ever more lavish. They reached their zenith in a recently published book.[1] In it, Kuklinski inflated his reported celebrity hit list to include Mafia boss Carmine Galante, shot to death in 1979, and another mob bigshot, Paul "Big Paul" Castellano, gunned down outside a steakhouse six years later. But nothing could top his controversial—and much disputed—claim

---

1. Philip Carlo, *Ice Man: Confessions of a Mafia Contract Killer* (New York: St. Martin's Press, 2006).

that he had been among a five-man gang of killers hired to get rid of Jimmy Hoffa. When the boss of the Teamsters Union went missing on July 30, 1975, it made for a sensational story. To date, his whereabouts remain a mystery. Enter Richard Kuklinski. In his version, Hoffa was lured from a Detroit restaurant, and after a short drive, Kuklinski knocked Hoffa unconscious with a blackjack, then plunged a hunting knife into the back of his head. Rather than dispose of the body locally, Kuklinski opted to drive it more than six hundred miles to a junkyard in Kearny, New Jersey, where it was placed in a fifty-gallon drum and set on fire. The drum, he claims, was later dug up and crushed. FBI sources scoff at Kuklinski's claims, dismissing them as the ravings of a swaggering solipsist. Others aren't so sure.

Unfortunately, Richard Kuklinski can no longer be questioned about any of these claims. On March 5, 2006, at seventy, the Iceman died at St. Francis Hospital in Trenton—of nothing more controversial than a cardiac arrest.

DISPUTED DOCUMENTS

The study of disputed documents essentially falls into two separate categories. The first deals with the provenance of handwriting—did such and such a person write a particular specimen of handwriting (note that disputed document analysis has nothing to do with graphology, which claims to discern personality through handwriting). The second deals with the actual document itself and is concerned with determining its veracity.

Probably all of us have noticed fluctuations in our handwriting, often from day to day. Any number of factors—mood, the position in which we sit, our physical well-being—can affect the way we put pen to paper. But to the accomplished observer, there is, beneath these superficial deviations, a distinctive style that comes through, defying any attempt at disguise. It might be the angle of writing, its uniformity across the page, or the manner in which letters are formed, such as whether letters like *g* and *h* are looped or not. Such analysis requires vast experience and a large investment of time before a considered opinion can be given.

With the advent of the typewriter, and more recently computer printers, the modern examiner must also be conversant with the history of print technology. Say, for example, that it could be shown that a document purportedly written in 1912 was typed on a machine that was not manufactured until 1938; then that document must be viewed with skepticism. Because the identifiable characteristics of each printing device increase with age as its mechanism wears and letters chip, the chances of another machine having identical defects becomes increasingly remote.

Other factors that have to be considered are ink and paper. Ink samples can be analyzed using thin-layer chromatography (see Explosives and Fire.) For comparison purposes, the Bureau of Alcohol, Tobacco, Firearms and Explosives has a reference collection of more than three thousand ink chromatograms. To overcome the risk of inaccurate ink dating, the bureau has suggested that manufacturers include a trace dye that changes annually. Another American agency, the Secret Service, maintains the International Ink Library, which it assumed in the 1920s from the police of the canton of Zurich, Switzerland. The library contains the chemical composition and other information, such as the date of formulation, on six thousand types of ink.

Determining whether a document has been altered is usually possible through use of an electrostatic detection apparatus (ESDA). A document is placed on top of an electronically charged metal mesh, and a thin plastic film is pulled tightly across it. As the document and the film are sucked tight onto the mesh, a mixture of photocopier toner and fine glass beads is applied. The mixture then clings to any electrostatically charged areas. When the original document is removed, all of the indentations on the film may be read. By matching all suspect original pages against each image, investigators can determine whether changes have been made to the original document.

Although primarily used in fraud investigations, disputed document analysis has played a significant role in all facets of crime detection, particularly homicide.

John Magnuson

DATE: 1922

LOCATION: Marshfield, Wisconsin

SIGNIFICANCE: From just a few scraps of bomb-damaged paper, investigators gleaned enough evidence to capture the Yule Bomb Killer.

A package mailed to the Marshfield home of James Chapman, chairman of the Wood County board of supervisors in central Wisconsin, on December 27, 1922, turned out not to be the belated Christmas present that was expected. As his wife, Clementine Chapman, eagerly tore at the wrapping paper, the parcel exploded in her face, fatally injuring her and maiming her husband.

It is a curious feature of bombs that most of the deadly blast radiates outward, often leaving the mechanism itself surprisingly intact. While the deadly device had carried out the deed for which it was designed—although it missed its intended target—scraps of paper from the packaging were still extant, and on those fragments could be seen faint traces of the handwritten address. These were turned over to John Tyrell of Milwaukee, Wisconsin's top examiner of questioned documents.

The postmark told investigators that the lethal package had been collected from the mailbox of Thorval Moen, who lived on Route 5 outside Marshfield. Moen denied all knowledge of the deadly parcel, pointing out that anyone could have left it in his mailbox, which was on a public highway. Careful checking of his alibi soon eliminated him as a suspect.

Tyrell, meanwhile, painstakingly reassembled the fragments of packaging. They read "J. A. Chapman, R. 1., Marsfilld [*sic*], Wis." From just these few words he would unlock the key to a killer's identity. The writing was so awkward that he first thought it had been deliberately disguised, but closer study of the spacing, slope, alignment, and pressure convinced him that it had been written to the best of the author's ability. Because literacy standards of the times were high, misspelling the town's name so badly suggested a reliance on phonetics rather than formal education. Rolling the pronunciation around on his tongue, Tyrell felt sure that the writer was foreign, almost certainly Swedish.

As it turned out, there was only one Swede in the community, John Magnuson, and he lived less than two miles from Moen's mailbox. It was no secret that Magnuson, a surly forty-four-year-old farmer, hated Chapman; for months the two men had been locked in a bitter feud over an intended drainage ditch that would bisect Magnuson's land. Dredges had been blown up and barns belonging to Chapman had been burned. With every incident the acrimony deepened.

Handwriting Sample

On December 30, officers visited Magnuson at his farm. In a barn they found particles of wood similar to the type used in the manufacture of the bomb. That same day, Magnuson was arrested and charged with first-degree murder. Unaware that the blast had not destroyed the wrapper of the fatal package, Magnuson had no compunction about supplying an example of his handwriting. At a glance, the similarities were apparent, although he did spell Marshfield as *Marsfild* instead of *Marsfilld,* as on the bomb.

Tyrell determined that the killer had used a medium smooth-pointed fountain pen with an unorthodox ink mixture: mostly Carter's black, but with a slight trace of Sanford's blue-black. When postal agents inspected Magnuson's farmhouse, they found that his daughter had a fountain pen with the very point Tyrell had described. Also, she always used Sanford's ink but had loaned her bottle to a schoolmate, who, when it ran dry, had refilled it with Carter's black ink, producing the exact mixture Tyrell had discovered on the death package.

To support Tyrell's findings, the prosecution enlisted the aid of a formidable team of handwriting experts, led by the redoubtable Albert S. Osborn of New York. He was joined by J. Fordyce Woods of Chicago. Both agreed that Magnuson had written the address on the package containing the bomb. Osborn noted fourteen distinct points of similarity, concluding that "by no coincidence could any two persons ever make the characteristic and peculiar repetitions as displayed on these documents."

An interesting lesson in phonetics was provided by J. H. Stromberg, professor of the Swedish Language Department at the University of Minnesota. Explaining idiosyncrasies in the Swedish language, he pointed out that the diphthong *sh* did not occur in the native tongue, and thus the word *marsh* would be pronounced *mars* and spelled that way, "especially by poorly educated Swedes." Similarly, the combinations *ie* and *ei* were also unknown in Swedish, leading an untutored immigrant to spell *field* as *filld* or *fild*. Such a person was likely to pronounce *Marshfield* as *Marsfilld* and write it that way.

There was other evidence against Magnuson. Arthur Koehler, the wood expert who would later achieve prominence during the Lindbergh kidnapping (see the Bruno Hauptmann case on page 302), matched fragments of white elm used in the bomb to shavings found on the floor of Magnuson's workshop. Metallurgist David E. Fahlberg of the University of Wisconsin described experiments comparing metal from the bomb's trigger with a sample of steel from Magnuson's workshop: "The two pieces of metal were examined and tested with chemicals in the process known as etching. Four separate tests were made by polishing with emery and with the chemical: it was found under a microscope which magnified one hundred diameters that the two pieces . . . are identical in pattern."

On March 31, 1923, Magnuson was found guilty. Two years later, Wisconsin's supreme court refused to set aside his life sentence.

Conclusion

In his final address, Magnuson's attorney, Charles Briere, fulminated against the "so-called experts" who had examined the scraps of the bomb, sneering that "half of them were here for their share of the gold bag of the state of Wisconsin. . . . The Almighty Dollar is what these men were after." It was a complaint about expert witnesses that echoes in courtrooms to the present day.

Arthur Perry

DATE: 1937

LOCATION: New York, New York

SIGNIFICANCE: So many factors were combined in this case that it has come to be regarded as an American detection classic.

Early on the morning of July 2, 1937, the battered body of a young woman was found on a vacant lot in the Jamaica district of South Queens. Beside the corpse, a baby girl lay crying but unharmed. The murder weapon—a bloodstained piece of concrete—was found some fifty feet away. Nearby, also caked in blood, were an electric iron and a man's black left shoe with a hole in its sole.

Initial theories that the woman had been the victim of a random attack while taking a shortcut through a deserted railroad underpass were soon discarded. If, as seemed likely, she had been carrying the baby when she was jumped, then how had the infant escaped injury? Immediately observers felt that the woman had been killed not by a stranger but by someone she knew and trusted. As crowds gathered, a woman stepped up who recognized the crying baby and was able to identify the victim as Phennie Perry, a twenty-year-old neighbor of hers who had recently moved.

The medical examiner fixed the time of death at not later than 2:30 A.M., and anywhere up to seven hours before that. A much closer estimate was provided by a night watchman from a local junkyard. Hearing screams at 10:10 P.M. the previous evening, he had called the police, but two patrol car officers, unimpressed by his drunken state, had left without investigating.

When the body was moved, a bundle of bloodstained papers was discovered. Rarely has a murder victim disclosed so many clues. As listed in the official record, the bundle included the following:

1. An envelope postmarked Trenton, New Jersey, Feb. 17, 1937, addressed to Mr. Ulysses Palm, 108–110 153rd St., Jamaica, with a letter inside addressed "Dear Son" and signed "Your mother, Marie Parham."
2. Another piece of paper addressed to Mr. Ulysses Palm.
3. A penny postcard addressed to Mr. Ulysses Palm, signed "Your niece, Ella Mae Parham."
4. A slip of paper addressed "Dear Member," and signed "C. K. Athetan."
5. An electric light bill bearing Palm's name and address.
6. Three photographs: a) a man in front of a car; b) a studio picture of the same man with a backdrop of a boat named SS *Leviathan;* c) a picture of a woman.
7. A church receipt book containing names and amounts of donations, receipts signed "J. Walker."
8. A bloodstained strip of blue broadcloth, about 3 inches long by half an inch wide.

Dual Suspects

When detectives visited Palm's address, they found that Phennie Perry and her husband, Arthur, a twenty-two-year-old construction worker, lived upstairs. But it was what they found in Palm's apartment that set pulses racing—the missing right black shoe and a bloodstained blue shirt with a piece torn off that matched the strip found under the body.

Both Palm and Perry were taken in for questioning. Palm, a deacon of the Amity Baptist Church, was obviously the man in the photographs found at the crime scene, yet he seemed genuinely perplexed as officers reeled off the damning catalog of clues found by the body. He readily admitted that most of the items—including the bloodstained shirt—were his, but he professed utter amazement as to how they had arrived at the crime scene. On one point he was adamant—the shoes belonged to Perry; he had given them to him sometime previously.

Across the hallway, Perry was recounting his version of the night's events. He told detectives that Phennie had met him after a bingo game at the Plaza Theater. Tearfully, she explained how Palm had tried to break into her bedroom that morning; she also produced an indecent letter Palm had written to her, which included a death threat if she showed it to her husband. Perry had stormed over to confront the lecherous deacon, and a bitter argument ensued. Palm denied all knowledge of the charge or the letter and demanded that Perry bring his wife to make the allegation in person. As Perry left, he noticed that the time was 9:50 P.M.; he had then called at the house of Phennie's sister, but his wife was not there.

The listening detectives shifted in their seats; either Perry had made a mistake, or he was lying through his teeth. They knew that Palm, who was a store clerk as well as a deacon, was nowhere near his home at 9:50 P.M. That night, for the first time in years, his employer had asked him to work late to conduct a stock inventory. Working together with several other employees, Palm did not leave the store in Flushing until 10:10 P.M. He had then taken a trolley home, a journey of about one hour. Therefore, at the time of the murder—almost certainly 10:10 P.M.—Ulysses Palm had a perfect alibi, which was more than could be said for Perry. An usher at the Plaza Theater claimed that she saw him leaving the premises at 10:00 P.M. with a woman and a baby.

Frame-Up

Palm admitted arguing with Perry, but said it happened after 11:15 P.M. Of course, genuine confusion could have arisen over the time, but the investigators doubted that. Gut instinct told them that Perry had killed his wife, then attempted to frame his neighbor. But to gain a conviction, prosecutors need evidence, not just suspicion. So the officers turned to New York's Technical Research Laboratory.

One of the investigators, Edward A. Fagan, began with Perry's clothing. It was hot in the room, and Fagan, noticing that Perry had his shirt sleeves hitched up past the elbows, asked him to roll them down. On the right sleeve was a tiny speck of blood. Also, one of Perry's socks had a blood-soaked patch on the sole that corresponded to the hole in the shoe found by the body. That same sock also revealed microscopic traces of earth, which Dr. Harry Schwartz matched to soil found at the crime scene.

It got better. Enlarged photographs of the torn blue shirt showed clearly that the tear had been started with a cut from a knife or scissors, then torn off by hand, to simulate a tear during a struggle.

The most damning evidence came from the threatening letter. Detective John A. Stevenson compared it to specimens of handwriting from Palm and Perry. While his expertise allowed him to eliminate Palm as the author, he felt unable to express an opinion as to Perry's involvement. Two of the nation's top handwriting experts, Elbridge W. Stein and Albert S. Osborn, were asked for their opinions. Their conclusions corresponded in every respect: Palm was entirely blameless; the letter had been written by Perry. Each man in his report highlighted certain peculiarities in the letters *u* and *b,* which were present in both the letter and Perry's writing sample, even though an attempt had been made to disguise the handwriting in the threatening letter. When a search of Perry's apartment uncovered sheets of writing paper that in every respect—size, thickness, weight, quality, and texture—matched the letter sent to Phennie, he was charged with first-degree murder.

Perry had planned every detail. In early June he stole the photographs from Palm's apartment. Then on June 20 he mailed the threatening letter, which set up the stage-managed argument with the hapless deacon. While Perry ranted, his wife was already dead and had been for an hour. Perry used the altercation to mask his theft of the receipt book. Later he returned to the crime scene and planted this further incriminating evidence on the body.

Sentenced to death after a five-day trial in November 1937, Perry then had his conviction overturned on a technicality. Retried a year later, he was again found guilty. On August 3, 1939, still protesting his innocence, Perry was executed.

Conclusion

It is unnerving to consider the possible outcome of this case had Palm not happened to work late that night. Without the twin interventions of fate and forensic science, it is entirely conceivable that he may well have taken Perry's place in the electric chair.

The Hitler Diaries

DATE: 1981

LOCATION: Hamburg, West Germany

SIGNIFICANCE: History's greatest publishing fraud was first legitimized, and then exposed, by scientific analysis.

On the morning of February 18, 1981, five men—all sworn to secrecy—gathered in the office of Manfred Fischer, director of the German publishing giant Gruner and Jahr, whose magazines included the best-selling *Stern*. Besides Fischer, there were other high-ranking company officials and one staff journalist, Gerd Heidemann. The reason for this conference lay on a table—three diaries, each bound in black and approximately one and a half centimeters thick. Handwritten in an almost illegible Germanic script, they represented the greatest publishing coup of the century—the diaries of Adolf Hitler.

Heidemann, who had located the diaries, was closemouthed about their source, other than to say that they came from a wealthy collector of Nazi memorabilia whose brother was a general in East Germany. By the meeting's end, without consulting a single document expert, scientist, or historian, Fischer committed his company to the purchase of twenty-seven volumes of Hitler's diaries at a price of eighty-five thousand marks each, plus a sum of two hundred thousand marks for the hitherto unsuspected third volume of Hitler's autobiography, *Mein Kampf*. At the 1981 exchange rate, this represented a total investment of approximately two million dollars.

Only later did Gruner and Jahr submit the diaries for authentication. Neither of the experts they chose was ideally qualified for such an undertaking, and neither could be blamed for the debacle that followed. Dr. Max Frei-Sulzer, former head of the Zurich police forensics department and a microbiologist by discipline, was asked to compare the script in the diaries with known examples of Hitler's handwriting. (Unfortunately for Frei-Sulzer, several of the purportedly genuine samples of Hitler's handwriting had emanated from the same source as the diaries!) He promised to report back.

They also submitted the documents to Ordway Hilton, whose credentials could not be disputed. A questioned-documents expert from Landrum, South Carolina, Hilton was top flight; Gruner and Jahr knew that his imprimatur on the veracity of the Hitler Diaries would greatly enhance sales prospects in the United States.

Nevertheless, there were serious flaws in their choice. Hilton did not understand Germanic script at all, and therefore he had to work from observation alone. Like Frei-Sulzer, he was unaware that the purportedly genuine samples of Hitler's handwriting were from the same source as the diaries. Given these disadvantages, his opinion that the documents were genuine was entirely excusable.

After two months' intensive study, Dr. Frei-Sulzer concurred: "There can be no doubt that both of these documents were written by Adolf Hitler." Further confirmation came when a handwriting expert from the Rhineland-Pfalz police department concluded "with a probability bordering on certainty" that at least three of the documents were genuine.

Millionaire Lifestyle

The experts' conclusions were just what Gruner and Jahr wanted to hear. Suitcases stuffed full of cash were handed over to Heidemann for onward transmission to his source. Oddly enough, no one at *Stern* appeared to notice that these transactions coincided with an astronomical rise in Heidemann's standard of living—he began stocking up on cars and real estate at a prodigious rate.

As the publication date drew near, Gruner and Jahr began putting out feelers in America and Britain, looking for partners. In the United States, both *Newsweek* and Bantam Books expressed an interest. Media tycoon Rupert Murdoch decided to bid for the world rights, and on April 9, 1983, he offered $3 million, the first salvo in a bidding war with *Newsweek* that would ultimately boost the price to $3.75 million.

Earlier that month Murdoch had flown the distinguished British historian Hugh Trevor-Roper to Zurich to examine the documents. This was a masterstroke. In matters pertaining to Adolf Hitler, Trevor-Roper's reputation was global and unmatched. He was, he admitted later, overwhelmed by the vast amount of documentation on display, and

Gruner and Jahr's insistence that three handwriting experts had declared the writing genuine. A trio of executives eagerly fielded all of Trevor-Roper's numerous queries about clear historical discrepancies. In the end, pressured by a consortium too heavily invested in the project to countenance the prospect of failure, Trevor-Roper declared himself satisfied that the diaries were authentic and set about writing an article to this effect, for publication in the Murdoch-owned London *Sunday Times*.

Well before this article appeared, West German police forensic scientists knew that the diaries were a hoax. At the request of Gruner and Jahr, they had conducted their own inquiry, and on March 28, 1983, Dr. Louis Werner filed a preliminary report of his team's findings. They had concentrated not on the handwriting, which had already been authenticated, but the actual paper used, to see if it stood the test of time. Werner was adamant: of the nine documents he had examined, at least six were forgeries. Under ultraviolet light all six appeared to contain a substance called *blankophor,* a paper-whitening agent that, as far as he knew, had not come into use until well after the Second World War. He was awaiting a second opinion, which came several weeks later.

Hoax

On May 6, 1983, German government scientists declared the diaries to be forgeries of the crudest kind. The paper—a poor-quality mixture of coniferous wood, grass, and foliage—had been treated with the chemical paper whitener blankophor, which had not come into use until after 1954. The book bindings also contained whitener, and threads attached to the official-looking seals were made from viscose and polyester, both modern products. Of the four types of ink used, none was known to have been widely used during the war; by measuring the evaporation of chloride from the ink, scientists established that the writing in the 1943 diary was less than twelve months old.

These revelations coincided with a crisis in Trevor-Roper's confidence about the wisdom of his article. His frantic attempt to stop publication failed by hours. When news of the hoax broke, Gruner and Jahr went looking for scapegoats. Gerd Heidemann, who throughout had been convinced of the diaries' veracity, faced double disaster—not only had he been duped, but millions of marks, originally earmarked for his source, had in fact wound up in his own bank account. Staring at a lengthy jail sentence, he at long last divulged the name of his source.

Far from being a wealthy collector, Konrad Kujau, forty-five, was the pettiest of criminals. In 1963 he had turned his dubious talents to the world of forgery, copying luncheon vouchers. From there he gravitated to forging paintings and Nazi memorabilia, and ultimately the Hitler Diaries.

On July 8, 1985, Heidemann was convicted of misappropriating his employer's money and sentenced to four years and eight months in prison. His erstwhile partner, Kujau, received four years and six months. The judge described *Stern*'s recklessness as tantamount to that of an accomplice in the hoax.

Following his release from prison in 1988, Kujau opened a gallery in Stuttgart, where he sold "genuine" forgeries of Hitler's paintings. No mere one-trick pony, he also turned out what he called "reinterpretations" of Dali, Monet, Rembrandt, and Van Gogh, signing them all with his own and the original artist's name. These became so popular that by the time of Kujau's death on September 12, 2000, a thriving market existed for his fakes.

What may—or may not—be the final twist in this bizarre saga came in April 2006, when Petra Kujau, a distant relative of Konrad's, was charged with forging her great-uncle's signature on fakes imported from Asia and then selling them on the Internet. Marketing these "fakes of fakes" had proved very profitable, allegedly netting her $680,000. Experts declared that they "weren't of very high quality, not comparable with the good forgeries made by Konrad Kujau."

Conclusion

In all, through outright swindle, royalties, fees, lost advertising and sundry other commitments, the Hitler Diaries were estimated to have cost *Stern* more than twenty million marks (sixteen million dollars). The cost in careers, reputations, and personal humiliation was incalculable.

Graham Backhouse

DATE: 1984

LOCATION: Horton, England

SIGNIFICANCE: This case provides an example of the interdependence of forensic disciplines that helps solve so many cases.

Among the rolling hills of England's West Country, Graham Backhouse and his wife Margaret had taken up farming in the small village of Horton. A previously successful hairdresser, Backhouse seemed by temperament and talent ill suited for the rigors of agricultural life, and so it proved. Two years of crop failures had spiraled his bank overdraft to an alarming seventy thousand pounds (more than one hundred thousand dollars) and had earned him nothing but contempt from his neighbors. They regarded the master of Widden Hall Farm as an arrogant interloper, and neither did they relish the sight of a forty-four-year-old married man cavorting so blatantly with numerous local women. The first indication that something more than mere village envy was at work came on March 30, 1984, when one of Backhouse's farmworkers found a sheep's head impaled on a fence. Attached was a handwritten note that warned: "You Next."

Backhouse stormed off to the local police station, where he detailed a lengthy and virulent campaign against him made up of threatening phone calls and poison-pen letters. Included among the calls, he said, was one accusing him of seducing the anonymous caller's sister and promising dire consequences if the relationship did not cease. Bereft of clues other than Backhouse's vague and petulant statement, the police could do little more than file the complaint and hope that the sheep's head episode had been an isolated incident. Unfortunately, such optimism proved groundless.

Explosion

On the morning of April 9, Backhouse asked Margaret to drive into town to pick up some medicine for the livestock. Inexplicably, her own car would not start, and after fruitless efforts to remedy the fault, she decided to take her husband's Volvo. When she turned the ignition key, a bomb hidden beneath the driver's seat erupted, ripping away half of Margaret Backhouse's thigh and causing countless lacerations to her body. She managed to crawl, screaming, from the smoking shell, but her cries went unheeded. Backhouse, some way off in a cowshed and with the radio at full volume, was oblivious to his wife's agony. Fortunately, the stricken woman was spotted by some school bus passengers and rushed to the hospital.

Combing the wreckage, explosives experts discovered that the bomb had been unsophisticated but deadly—a length of steel pipe packed with nitroglycerine and shotgun pellets wired to the ignition. Luckily, most of the explosive force had been deflected downward by the driver's seat. Even so, doctors had to remove thousands of pellets and fragments of shrapnel from Margaret's body. While she was still in the hospital recovering from her ordeal, Backhouse received yet another threatening letter.

This communication was sent to document examiner Mike Hall at the Birmingham Forensic Laboratory. He quickly realized that any attempt to identify the handwriting was pointless, as each letter had been gone over backward and forward repeatedly in order to disguise the hand. But when he inspected the "You Next" note, he came up with something of interest. On the back, barely visible, was the faint impression of a doodle, probably made on an adjoining sheet in a notepad. It wasn't much, but Hall carefully stored the note for later reference.

Meanwhile, Backhouse had been talking to the police. Clearly shaken by the incident—for he was convinced that the bomb had been intended for him—he maintained that he had no enemies, an odd remark considering his earlier complaint about the hate campaign. When pressed on this inconsistency, Backhouse reluctantly conceded that a neighbor, Colyn Bedale-Taylor, might possibly bear a grudge. The two men had been engaged in a lengthy and violent dispute over a right-of-way, an argument exacerbated by Bedale-Taylor's severe depression since the death of his son Digby in a car crash two years previously. Warming to his theme, Backhouse suggested another possible suspect. This person worked in a quarry and therefore had access to explosives. Questioned as to why this particular fellow might wish him harm, Backhouse admitted to having had an affair with the man's wife.

For his own safety, Backhouse was given police protection at Widden Hall Farm. This arrangement ended after just nine days, when Backhouse angrily phoned Detective Chief Inspector Peter Brock and ordered him to remove the officers from his land. Backhouse insisted that he was quite able to defend himself. Aware that Backhouse owned a shotgun, Brock warned him against taking the law into his own hands, then withdrew as requested. As a compromise, Backhouse agreed to the placement of an alarm button connecting the farmhouse directly to the local police station.

For two weeks, all was quiet at Widden Hall Farm, but on the evening of April 30, the alarm shrilled. Police rushed over. Just inside the front doorway, they found a scene of appalling mayhem. Backhouse, barely able to stand, was awash in blood, his face slashed several times down one side. Another, deeper gash ran from his left shoulder diagonally across his body. At his feet lay sixty-three-year-old Colyn Bedale-Taylor, dead from two shotgun blasts to the chest, his lifeless fingers wrapped around a Stanley knife.

Backhouse told police that the dead man had called to inquire after Margaret's health, but had suddenly gone berserk. Crying out that

God had sent him, Bedale-Taylor admitted planting the bomb, but said it was because Backhouse had caused the death of his son. He then lunged at Backhouse with the Stanley knife, slashing him about the face. Backhouse fought off his demented neighbor and fled down the hall. "I ran into the hallway and grabbed a gun. Bedale-Taylor was still after me. I shouted I had got a gun, but he still kept coming and I shot him. He fell back and I shot him again and that was it."

Strange Engraving

Given recent events, it sounded like a plausible enough story: The Stanley knife bore the initials *BT,* and Bedale-Taylor's mental confusion was well known. Further corroboration came when officers searched the dead man's house and found the actual steel pipe from which the car bomb had been cut. But they also noticed something else. All of the tools in Bedale-Taylor's workshop—he had been a carpenter—were carefully engraved *BT,* unlike the crudely monogrammed Stanley knife. They found it curious, to say the least.

Experts who studied the crime scene were also perplexed. Dr. Geoffrey Robinson, a biologist, pointed out that the drops of blood on the kitchen floor were of the wrong shape. During a frenzied life-and-death struggle, blood splashes typically land with distinctive tails; yet these spots were uniform and round, as if they had dripped from someone standing motionless. Other factors didn't add up either. There was no trail of blood from the kitchen to the hall, the direction in which Backhouse said he had run for his life, and neither did he display any of the usual defensive wounds to his hands, customary when someone has attempted to ward off a knife attack.

Additional suspicion was cast by pathologist Dr. William Kennard, who expressed doubt that the Stanley knife would still be clasped in the dead man's hand and not lying on the floor. The only situation in which he could imagine such an occurrence was if rigor mortis had formed immediately, a physical impossibility. Kennard postulated that Backhouse had first shot Bedale-Taylor and then put the knife in his hand.

Even more incriminating was Robinson's examination of the envelope that had contained the first threatening letter sent to Backhouse. Beneath the gummed flap he found a tiny wool fiber. Microscopic analysis matched this with Backhouse's own sweater, strongly suggesting that he had mailed the letter to himself. Suspicion turned to certainty when detectives searched Widden Hall Farm and found a notebook in a drawer. Inside was the page that still bore the original doodle found impressed on the "You Next" note. Enlarged photographs of the impression superimposed over the doodle provided an exact match.

Even while in prison awaiting trial for murder, Backhouse continued plotting. He persuaded a fellow prisoner to smuggle out an unsigned letter to a local newspaper implicating Bedale-Taylor in the car bombing. The scheme backfired hopelessly when the handwriting in this letter was analyzed and found to be indisputably that of the prisoner.

Conclusion

Piece by piece, the magnitude of Backhouse's fiendishness became apparent. In early March, he had increased the insurance on Margaret's life from fifty to a hundred thousand pounds, waited a few weeks while spreading word of a nonexistent hate campaign, then planted the bomb that so nearly killed her. When that attempt failed, and to divert suspicion from himself, he had lured Bedale-Taylor to his house with the sole intention of killing him. The seriousness of his self-inflicted wounds almost fooled the authorities, but he had underestimated the astonishing scope of modern forensic detection. On February 18, 1985, Backhouse learned the price he would have to pay for that arrogance—two terms of life imprisonment.

DNA TYPING

Even though the discovery of chemical DNA can be dated to 1869, it was the Russian-born biochemist Phoebus Levene, in 1911, who first discerned that individual cells each contain a nucleus made of nucleic acid. There are two types of nucleic acid, which he called ribonucleic acid (RNA) and deoxyribonucleic acid (DNA), according to whether they contain ribose or deoxyribose. Within each nucleus are twenty-three pairs of chromosomes made up of DNA. Within each pair, one chromosome comes from the father's sperm, the other from the mother's egg. By the 1940s, scientists realized that DNA forms the building blocks of life and dictates not only our hair and eye color, but everything about our physical makeup.

So how does DNA issue its instructions? Scientists knew it had to be in some kind of code and that every code would be unique to each individual. The answer came in 1953, when two scientists based in Cambridge, England—James Watson and Francis Crick—determined that the structure of DNA is a double-helix polymer, a spiral consisting of two DNA strands wound around each other. Biophysicist Maurice Wilkins's work on X-ray diffraction had proved crucial to Crick and Watson; all three men were awarded the 1962 Nobel Prize in Physiology or Medicine for their work in this field.

Four chemicals make up DNA: adenine (A), guanine (G), cytosine (C), and thymine (T). Strung together in an extremely long sequence, the chemicals on one chromosomal strand always align with the chemicals on the other strand; that is, A always joins with T, and C always joins with G. A section of DNA code might be arranged thus:

C A G T T C A

G T C A A G T

Although large chunks of DNA are universal (because each of us has the same body parts and organs), some sections of DNA vary from individual to individual. By studying these *polymorphic* segments, scientists can determine whether a particular strand of DNA could have come from a given individual. By comparing prints of several different polymorphic sequences from different specimens to one another, scientists can tell whether the specimens match.

The procedure for creating a DNA fingerprint in a criminal investigation is shown in the following steps (the sample used is blood, but it could be any bodily fluid or tissue):

1. Blood samples are collected from the victim, defendant, and crime scene.
2. White blood cells are separated from red blood cells.
3. DNA is extracted from the nuclei of white blood cells.
4. A restrictive enzyme is used to cut fragments of the DNA strand.
5. DNA fragments are put into a bed of gel with electrodes at either end.
6. Electric current sorts DNA fragments by length.
7. An absorbent blotter soaks up the imprint; it is radioactively treated, and an X-ray photograph (called an autoradiograph) is produced.

Once the autoradiograph has been analyzed and an apparent match found, the question is one of probabilities: What is the statistical likelihood of two people sharing this DNA profile? According to Sir Alec Jeffreys (see the Colin Pitchfork case on page 71), the answer is fewer than 1 in 1 nonillion (1 followed by thirty zeroes)—a figure billions of times greater than the current world population.

If only a small amount of DNA is available for fingerprinting, a polymerase chain reaction (PCR) may be used to create thousands of copies of a DNA segment quickly and accurately. Developed in 1983 by the Nobel Prize–winning American biochemist Kary B. Mullis, PCR allows the investigator to obtain the large quantities of DNA necessary for high-quality forensic analysis. It is a three-step process carried out in repeated cycles, and requires as little as a single DNA molecule to serve as a template.

The initial step—denaturation or separation of the two strands of the DNA molecule—is achieved by heating the sample to around 95°C (203°F).

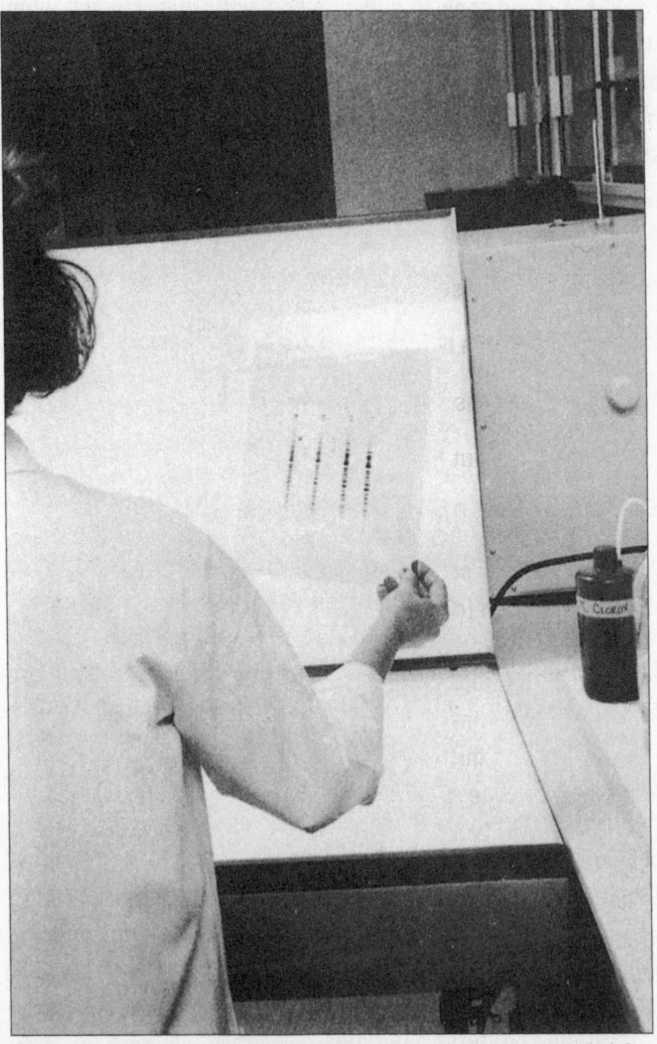

DNA—the future of forensic science. Here a laboratory technician reviews a DNA autoradiograph. (FBI)

Each strand is a template on which a new strand is built. Steps two and three involve cooling and reheating the sample, a process that doubles the number of copies. At the end of the cycle, which lasts about five minutes, the process begins again. Usually 25 to 30 cycles produce a sufficient amount of DNA.

Despite teething problems with overzealous laboratories scrimping on their test procedures in order to produce quick and profitable results, DNA profiling has survived the most rigorous courtroom scrutiny and continues to prosper. With its unparalleled ability not only to convict the guilty but to free the innocent—by May 2006 in the United States, no fewer than 175 wrongfully convicted prisoners had been exonerated—it is the greatest advance in forensic science since the advent of fingerprinting.

The Romanovs

DATE: 1918

LOCATION: Yekaterinburg, Russia

SIGNIFICANCE: One of the twentieth century's greatest mysteries was finally solved by DNA typing.

History records that on July 17, 1918, Czar Nicholas II of Russia; his wife, Alexandra; their five children; and four members of their entourage were executed in the cellar of a house in Yekaterinburg, Siberia, on orders of Lenin, the newly instituted leader of the Bolshevik revolution. Their death brought an end to the three-hundred-year reign of the Romanovs and touched off worldwide interest in their fate. Although for decades Soviet governments forbade all mention of the deposed royal family, some were prepared to flout the law. The fateful house where they were executed became a shrine, with pilgrimages so commonplace that in 1977 it was bulldozed to the ground.

For one man, Gely Ryabov, finding the burial site of the Romanovs became an obsession. Using his privileged position as a filmmaker for the Interior Ministry, he delved into secret archives and in the late 1970s managed to track down the children of Yakov Yurovsky, the Bolshevik guard who had overseen the executions. Yurovsky's son provided Ryabov with a hitherto unknown note that described the disposal of the bodies in a swampy meadow near Yekaterinburg. With its help, Ryabov located the layer of logs shallowly covered with dirt that lay over the muddy spot where the remains were buried. A local historian helped him, along with a geologist who scaled a pine tree to spot traces of the old road traveled by the truck that carried the corpses.

Ryabov, aware that such activities would bring censure at best and probably much worse, carried out most of his work at night, but eventually he uncovered a pile of black and green bones that he felt sure were those of the former czar and his family. Also recovered from the burial site were scraps of expensive clothing that seemed to correspond roughly in gender, age, and size to the Romanov family and entourage.

As the communist stranglehold weakened in 1991, the work to establish the identity of the corpses began. Using computer superimposition software, Aleksandr Blokhin and other researchers first compared the battered skulls with photographs of the czar and czarina. Initial results suggested that these were indeed the Romanovs. But for a positive identification they turned to DNA typing.

Advanced DNA

The problem was especially daunting because the scientists had only bone to work with. Unlike living tissue or vital fluids, bones contain very little DNA, and these bones were in particularly bad condition. Some were so fragile that they turned to dust when touched. One of Russia's foremost DNA specialists, Pavel Ivanov, took the bones to Britain for further study by the Home Office Forensic Science Service. Conventional DNA analysis established that five of the nine skeletons were members of one family—a man, a woman, and three children. But because of their age and deteriorated condition, the bones did not provide scientists with all that they needed to know.

They turned instead to mitochondrial DNA, which is found in specific structures within the body's cells and is passed down intact through the maternal line. The forensic team asked for and received a blood sample of Queen Elizabeth's husband, Prince Philip, who is a direct descendant of Czarina Alexandra's sister. If the bones were genuine, the team said, then Prince Philip's DNA should match that of the woman and the three children in certain familial respects. The results enabled chief researcher Peter Gill to announce a "complete match"; the bones' provenance had been established "virtually beyond doubt." Gill declared himself and his team "more than 98.5 percent certain that the remains are those of the Romanovs."

However, one mystery endured. History records that eleven people were executed in the cellar, yet only nine corpses were recovered. Yurovsky's note tells of burning two other bodies but gives no reason why. Curiously enough, the two bodies not found were those of the Crown Prince Alexei and his sister Anastasia, both of whom were rumored to have survived the assassination. In 1920, a woman who became known as Anna Anderson surfaced in Berlin, claiming to be Anastasia. Over the years she convinced many people, including some distant members of the royal family, of her claims, but she was never officially regarded as the missing princess. Others regarded her as a fraud. She died in Charlottesville, Virginia, in 1964, still sticking to her story. Nothing found at the burial site appeared to refute her claim.

But in 1994, the truth became known. Before her death, Anna Anderson had undergone an operation. To provide legal cover in case of a possible compensation claim, the hospital had retained a sample of tissue. Supporters of Anderson commissioned Gill to test the sample and compare it with the Romanovs' DNA. In June 1994, Gill flew to the United States in secret to collect the sample. His findings, released the following October, exploded the myth. DNA typing proved that Anna Anderson was an impostor. She was not Princess Anastasia but a neurotic Polish peasant named Franzisca Schanzkowska who had vanished in 1920. Later that same year, she was pulled from a canal in Berlin and began her long-standing deception. Members of the Schanzkowska family, who had long-suspected that Anna Anderson was their missing relative, provided the samples that fixed her identity.

Conclusion

The outcome of this investigation has still not satisfied everyone. Resolute supporters of Anna Anderson, oblivious to the evidence of her genes, continue to insist that she was the missing princess.

Colin Pitchfork

DATE: 1983

LOCATION: Narborough, England

SIGNIFICANCE: This landmark case can claim a double triumph—the first suspect eliminated by genetic fingerprinting and the first murderer caught by it.

One of Britain's longest and most exhaustive manhunts began on an icy winter day in late 1983. It was twenty past seven on the morning of November 22 when a hospital porter making his way to work in the village of Narborough, near Leicester, took his usual shortcut along a lonely footpath known as the Black Pad. A few yards along the lane, he saw, sprawled on some grass and white with frost, the body of fifteen-year-old Lynda Mann. Nude from the waist down, she had been strangled and raped the previous evening while on her way to visit a friend. Semen traces showed that the killer was a Group A secretor (see the Joseph Williams case on page 249) with a strong phosphoglucomutase (PGM) 1+ enzyme. Taken together, these two factors occur in only 10 percent of the male adult population, a scarcity that heartened detectives. Although this could not positively identify the killer, it would certainly reduce their list of possible suspects.

Initial inquiries centered on Carlton Hayes Hospital, a nearby mental institution. When that failed to bear fruit, the search radiated out to include the adjacent villages of Enderby and Littlethorpe. Despite their conviction that the killer was a local man, investigators canvassing the three-village area were stymied at every turn. Only later would the awful realization sink in that they had actually questioned the killer during this sweep.

The computer had flagged the man for two reasons: (1) he had previous convictions for indecent exposure, and (2) he had been referred for therapy as an outpatient at Carlton Hayes Hospital. Although unable to provide an alibi—he claimed to have been taking care of his son—the fact that at the time of the murder he had lived a few miles beyond what police regarded as the probable catchment area outweighed all other considerations, and he was eliminated from the inquiry. (He did not move to Littlethorpe until a month after the killing of Lynda Mann.) Also, the likelihood of a parent taking time off from child care duties to commit such a horrendous murder was thought too grotesque to countenance.

Gradually the investigation into the murder of Lynda Mann ran out of steam, stifled by a lack of clues and diminishing public interest. On the first anniversary of Lynda's death, someone left a tiny cross at the spot where her body was found. A year later the commemorative ritual was repeated. Before it could happen a third time, the killer struck again.

Second Murder

Dawn Ashworth was also fifteen. A schoolgirl from Enderby, she disappeared in broad daylight on the afternoon of July 31, 1986. Two days later her hideously violated body was found less than a mile from the spot where Lynda Mann met her death. She had been torn to bits. Semen tests confirmed that detectives were hunting a dual killer. In the aftermath of this murder, investigators got their first real lead. A kitchen porter at Carlton Hayes Hospital, a slow-witted youth of seventeen, seemed to know an awful lot about the killing of Dawn Ashworth. Even though a blood test showed he was not a PGM 1+, Group A secretor, his confession had the ring of authenticity to it, especially to the officers who had spent almost three years tracking Lynda Mann's killer, and despite the lad's troubling insistence that he knew nothing about the first murder.

Just about the only person who believed that the boy knew nothing about either killing was the suspect's own father. He wondered if the magazine article he had recently read about a thirty-six-year-old scientist at nearby Leicester University who had perfected a system of identification based on DNA called genetic fingerprinting would clarify matters.

In the autumn of 1984, Dr. Alec Jeffreys, a research fellow at Leicester University's Lister Institute, stumbled upon the discovery that would play a pivotal role in this and countless subsequent cases. Although the existence of DNA had been known for decades, this unassuming scientist perfected the means whereby identifiable genetic markers could be developed on an X-ray film as a kind of bar code and then compared with other specimens. Although accounts vary of how Jeffreys actually became involved in the murder investigation, he was eventually asked to extract DNA from the killer's semen and compare it to the kitchen porter's blood sample. His results stunned those leading the investigation. Not only had the porter not killed Lynda Mann, but he hadn't killed Dawn Ashworth, either! His entire confession had been a fabrication! Just about the only good news that Jeffreys had to offer the jaded officers was confirmation that one man had killed both girls. On November 21, 1986, legal and forensic history was made when the teenaged kitchen porter became the first accused murderer to be set free as a result of DNA fingerprinting.

For those charged with finding the double killer, the verdict had wider ramifications—if DNA typing was so accurate, then why not conduct a mass testing of the local male population? After all, there were precedents for such action (see the Peter Griffiths case on page 127).

Early in 1987, investigators decided to draw blood from every local male between the ages of sixteen and thirty-four for DNA testing. By the end of January one thousand men had been tested but only a quarter had been cleared, because the laboratories were overwhelmed by samples. (In its original form, DNA typing was a laborious, time-consuming procedure, often taking weeks. The process has now been reduced to a matter of days.) It was the same the following month: hundreds more tests, but no clues. Of course, the police did not expect the killer to volunteer blood, but they were hoping to flush him out. Those who refused to cooperate with the official request soon found themselves under the most intense scrutiny. It was a war of nerves that paid off in the strangest way.

Conspiracy

On August 1, 1987, a quartet of drinkers in a Leicester pub, all bakery workers, were discussing the notorious sexual liaisons of a fellow employee named Colin Pitchfork, when one of the four, Ian Kelly, dropped a conversational thunderbolt—Colin, he said, had bullied him into taking the blood test on his behalf. A deathly hush fell over the table. It was broken at last by another man present, who chimed in that Pitchfork had approached him also, offering two hundred pounds (three hundred dollars) if he would act as a stand-in, but he had declined. Pitchfork had told both men that he was scared to take the test because his record—he had convictions for indecent exposure—meant that the police would give him a hard time. Kelly, a timid, malleable person, finally caved in under Pitchfork's relentless pressure and, using a faked passport, had gone along and given blood in Pitchfork's name.

A woman sitting at the table listened to these revelations with an anxiety born of suspicion. Like everyone else who worked at the bakery, she knew Pitchfork as an overbearing lecher, forever harassing the female employees. But did that make him a murderer? And then there was Kelly and possibly the other man to consider. If she went to the authorities, what would happen to them? For six weeks she wrestled with her conscience, then contacted the police. First, detectives arrested Kelly; later that day, September 19, 1987, they called at the Littlethorpe home of twenty-seven-year-old Colin Pitchfork. He took his arrest philosophically and made no attempt to deny either killing.

For those who had staked their reputations on the efficacy of DNA testing, this would be the acid test. A sample of Pitchfork's blood was rushed to Jeffreys's laboratory. After painstaking examination, the genetic bar code was found to be identical to that of the DNA sample from the killer-rapist. Colin Pitchfork was the 4,583rd male to be tested, and the last. The principle had been vindicated.

On January 22, 1988, Pitchfork pleaded guilty to both murders and was jailed for life. The judge, recognizing Ian Kelly's role in the affair as that of gullible pawn, gave the hapless bakery worker an eighteen-month suspended sentence.

Conclusion

This verdict reverberated around the world. Within a year of Pitchfork's conviction, American DNA laboratories had been consulted in more than one thousand criminal cases. For Dr. Alec Jeffreys the accolades outweighed the income. He never made any fortunes from his invention, but in 1994 the quiet professor from Leicester did receive a knighthood.

Kirk Bloodsworth

DATE: 1984

LOCATION: Baltimore, Maryland

SIGNIFICANCE: After a convicted killer had spent a year on death row and almost a decade behind bars, his fate rested in the hands of DNA typing.

On July 24, 1984, Dawn Hamilton, nine, was found raped and bludgeoned to death with a rock in a wooded area near her home in Rosedale, a suburb of Baltimore. Two weeks later, police arrested local resident Kirk Bloodsworth after receiving an anonymous tip from a caller who said that the twenty-three-year-old Eastern Shore waterman resembled a composite sketch circulated in local newspapers. According to prosecutors, Bloodsworth had returned to his hometown of Cambridge, some one hundred miles south of Baltimore, shortly after the slaying and told acquaintances he had "done something terrible" that could harm his marriage. Bloodsworth, an ex-Marine with no previous criminal record, insisted that he had been referring to his marijuana use and the anguish it had caused his wife.

At his first trial in 1985, despite alibi witnesses placing Bloodsworth at home or with other people during most of the day of the killing, he was found guilty of murder and sentenced to death. A year later, the Maryland Court of Appeals overturned the sentence and ordered a new trial because, it said, police had failed to inform Bloodsworth's attorneys at the time about another suspect in the case, a newspaper deliveryman who was seen in the woods just before Dawn's body was discovered. Allegedly, this man's shirt was stained by a spot that might have been blood.

Bloodsworth fared little better at this second trial. He was convicted again on the same charges, but this time the judge sentenced him to three terms of life imprisonment.

In all the time he spent behind bars, Bloodsworth maintained his innocence. Finally, in 1992, his attorney, Robert E. Morin, was granted permission to reexamine the physical evidence in the case. Using forensic techniques developed since the original investigation, a spot of semen less than one-sixteenth of an inch wide was discovered on the victim's panties. In handing over the evidence to Morin, the prosecution signed a letter stating they would "agree to [Bloodsworth's] release" if a laboratory ever "determines with scientific certainty" that any sperm found did not belong to Bloodsworth and if prosecutors confirmed the results independently. This final clause was critical. Much of the controversy that has surrounded DNA typing has concerned the quality of testing (some laboratories, in their haste to grab a share of the lucrative DNA market, have done less than exemplary work). Maryland state prosecutors, quite reasonably, wished to have the tests performed by two independent sources, one of their own choosing.

Innocent

A California laboratory hired by the defense, Forensic Science Associates, determined that the DNA of the semen did not match a sample of blood provided by Bloodsworth. Their report concluded: "He has been eliminated as a potential source of the sperm from the panties." The second test was carried out by America's most sophisticated facility—the FBI Laboratory in Washington, D.C. Their scientists also confirmed that the sample exonerated Bloodsworth. In the face of these results, and mindful of her office's earlier promise, Baltimore County state's attorney Sandra A. O'Connor formally moved to dismiss the charges against Bloodsworth in 1993, saying that there was "insufficient evidence" to mount a new trial. O'Connor, known as a tough prosecutor and an advocate of capital punishment, pointedly refrained from saying that she believed Bloodsworth innocent; she offered no apologies. "I am saying there is not enough evidence to convict him." She admitted that had investigators had the DNA test available to them in 1984 or 1987, "we would not have charged him. We would have continued the investigation." She concluded the statement by saying there is an "ongoing investigation" of the case, but she said that "there are no suspects at this time."

Released on June 28, 1993, after nine years in prison, Bloodsworth announced his intention to remarry. He was officially pardoned on December 22, 1993.

Conclusion

At the time of Bloodsworth's release, Peter Neufeld, chairman of the National Association of Criminal Defense Lawyers' DNA task force, said that nationally about a dozen inmates had been freed because of DNA testing. The list is still growing.

Tommie Lee Andrews

DATE: 1986

LOCATION: Orlando, Florida

SIGNIFICANCE: This case ended with the first American trial to admit DNA typing into evidence.

On May 9, 1986, Nancy Hodge, a twenty-seven-year-old computer operator at Disney World, was in the bathroom of her apartment when she heard a noise behind her. She turned to see a stranger looming in the doorway, knife in hand. In the brief violent struggle, she caught a glimpse of his face before being thrown to the floor. At knifepoint she was raped three times. Throughout the ordeal, the attacker covered her face so she could not see him. Afterward, he grabbed her purse and fled.

Over the following months, the rapist honed his strategy. Always he made sure that the victim could not identify him, usually by covering her head with a sheet or blanket. Then he would switch on the light as a prelude to the assault. Another quirk was to invariably take away some item belonging to the victim, often a driver's license. By December 1986, Orlando detectives were hunting a serial rapist who had struck no fewer than twenty-three times in the southeastern part of the city that year.

The rapist's pattern served him brutally well until February 22, 1987. He broke into a twenty-seven-year-old woman's home in the usual way, through a window, then beat and slashed her into submission with a knife. Half-smothered with a sleeping bag, the woman stifled her cries rather than wake her two children, who were asleep in the next room. But on this night, the rapist slipped up: Experts found two fingerprints on the window screen.

The police tightened their surveillance cordon around the neighborhoods where the attacker struck most. Their diligence paid off in the early hours of March 1, 1987, when a woman called to report a prowler. Within minutes, patrol cars were at the scene, just in time to see a blue 1979 Ford racing away. They chased the car for two miles until the Ford spun out of control on a bend and crashed. The driver, twenty-four-year-old Tommie Lee Andrews, worked at a local pharmaceutical warehouse. His fingerprints matched those on the window screen, and he was charged with raping the young mother a week earlier.

But the police wanted more: If they could prove that Andrews was a serial rapist, he would face life imprisonment; a onetime offender might be walking the streets again in a few years. At a police lineup, Nancy Hodge unhesitatingly identified Andrews as the man who had raped her. Yet in spite of a clear identification, the case was far from watertight. Of all the rape victims, only Nancy Hodge had seen her attacker's face, and then just for a few seconds. Eyewitness testimony is notoriously fallible. Not even the fact that Andrews's blood group matched semen samples taken from the victims could be considered significant—so did that of 30 percent of the male population.

Pioneering Technique

While the case was still waiting to go to court in August 1987, assistant state's attorney Tim Berry heard about British successes with DNA fingerprinting and wondered if it might provide the conclusive evidence he was seeking. He contacted Michael Baird, forensic director for a New York–based company called Lifecodes, which had begun DNA testing in America. Baird requested samples of the rapist's semen and blood from the accused man. The tests were carried out by Dr. Alan Giusti. In early October, the results came back: The bar codes from the two sources were identical. Beyond any reasonable doubt, both blood and semen came from the same man.

Now came the hard part—convincing a court that DNA typing was scientifically sound and that it was admissible as evidence. After a lengthy pretrial hearing, the judge agreed that it was, and on October 27, 1987, Andrews stood accused of raping Nancy Hodge. On the witness stand, she identified Andrews as her assailant. His defense—that he had spent that whole evening at home—was supported by his girlfriend and his sister. In submitting the DNA evidence, the prosecutor stated that the odds against Andrews being falsely accused were one in ten billion. It sounded impressive enough until the defense challenged him to substantiate this figure. Caught without the necessary data, the prosecution was forced into a humiliating retraction. How much it affected the jury's inability to reach a verdict is unknown, but the split—eleven to one in favor of conviction—was irrevocable and resulted in a mistrial.

Two weeks later, Andrews stood trial on the second rape charge, and this time the prosecution was fully prepared for any eventuality. Furthermore, they had fingerprint evidence to back up their DNA claims. To no one's surprise, Andrews was found guilty and sentenced to twenty-two years.

At the retrial of the Hodge case in February 1988, the DNA evidence assumed even greater significance. Andrews's alibi was still intact, verified by his sister and girlfriend. Nancy Hodge was equally emphatic in her identification. So everything hinged on the DNA. Using graphs and charts, Baird and David Houseman, a research biologist at the Massachusetts Institute of Technology, testified that the DNA in the blood and the semen matched. Baird responded to defense lawyer Hal Uhrig's charge that the test was invalid because not all the DNA molecules had been analyzed by explaining that it was necessary to analyze only those parts of the molecule that differ from person to person, not the characteristics found universally. The jury absorbed all of this and on February 5, 1988, found Andrews guilty. As a convicted serial rapist, he received jail terms totaling 115 years.

Conclusion

After a spluttering start, DNA evidence has now been accepted in courts across the country. There are still challenges to the discipline, chiefly on the grounds of shoddy laboratory work, but thus far, no one has successfully undermined the basic principle. Until that happens, DNA is here to stay.

Ian Simms

DATE: 1988
LOCATION: Billinge, England
SIGNIFICANCE: This case established the principle of "DNA by proxy."

With a brutal winter storm lashing around her, Helen McCourt alighted from the bus in Billinge, a village to the northeast of Liverpool, and steeled herself for the ten-minute walk home. It was a journey she never completed. For reasons known only to herself, the twenty-two-year-old insurance clerk stopped on the way at a pub called the George and Dragon. Like several local girls, Helen had fallen prey to the proprietor's brand of swaggering machismo. But just lately, their affair had soured, especially since they had had some trouble in the bar a couple of nights earlier, when she had ended up fighting with another woman. On that occasion, the proprietor, Ian Simms, had been overheard saying how much he hated his former lover. Tonight—February 9, 1988—Helen McCourt was on a retrieval mission. As she climbed the hill, the pub stood in darkness, but she knew that Simms would be in the upstairs apartment; he always was at this time. No one saw her enter the George and Dragon, and no one saw her leave. In fact, from that day to this, no one has ever admitted seeing Helen McCourt again.

At eight o'clock that night, her mother, frantic with worry, called the police. In such a close-knit community, news of Helen's disappearance spread like wildfire, and so did gossip concerning an alleged scream—which led police to Ian Simms. When officers came face-to-face with the husky bodybuilder, they were struck by his extreme nervousness—he was so nervous he could hardly speak. And what he did say reeked of hasty and inadequate fabrication. The investigators found mud on his bracelet and on two rings, for which Simms had no plausible explanation. Nor could he account for the two visible scratches on his throat, other than to hurriedly blame them on an argument with his wife. "Then what about the reported scream coming from the direction of your pub at around 5:30 P.M. on the night Helen disappeared?" he was asked. Simms mumbled that he knew nothing about it.

Detective Superintendent Thomas Davies, in charge of the investigation, thought otherwise and impounded Simms's Volkswagen. In the trunk were a bloodstained opal and sapphire earring, later identified as belonging to Helen McCourt, and also eighteen-inch-long strands of hair. Searchers moved upstairs to the apartment. One of the first items they found was a clip from the earring. And there was blood everywhere—enough for forensic scientist Dr. Ian Moore to map out Helen McCourt's last moments alive. The fight had started just inside the door to Simms's apartment. A trail of blood led upstairs, across the landing to a rear bedroom, where splashes on the walls showed how Simms had beaten her to the ground and then had continued his attack at floor level. At some point, he murdered Helen, by either strangulation or pummeling her to death, possibly through stabbing. Afterward, he had bundled her body into the trunk of his car.

Sex Assignation

Late that night, after the pub had closed, with the body of Helen McCourt lying in the trunk of his car, Simms drove the five hundred yards to his home to see his wife and two children before returning to the pub, where he had sex with yet another young admirer. When this girl left at 1 A.M., Simms was then believed to have dumped the body. But where? There was still no sign of Helen McCourt's corpse.

Despite the absence of a body, Davies felt he had enough to prove a charge of murder, and Simms was taken into custody on February 14. The evidence began to pile up remorselessly. Just over two weeks later, a marksman out shooting rats found a woman's purse in a field at Irlam, about seventeen miles from Billinge. It was identified as belonging to Helen McCourt. Detectives extended their search to Hollins Green, and there, on the banks of the River Irwell, they recovered Helen's coat and some items of clothing belonging to Simms, all heavily bloodstained, as well as a knotted length of electric cord. Strands of hair entwined in the cord matched those found in the trunk of Simms's car, leading investigators to conclude that this had been the ligature that choked the life from Helen McCourt. Among the items of clothing was a sweatshirt bearing the logo of the Labatt brewing company, which coincidentally had been running a promotional campaign at the George and Dragon.

Detectives steadily assembled a vast forensic battery against Simms, more than nine hundred samples. Dr. Moore tested the hair recovered from the car and the electric cord and reported that "there was an unusual color gradation along the length of them, and they matched hairs recovered from Helen's bedroom." Plastic trash bags, used to hold the bloodstained clothes, were matched to bags at the George and Dragon by comparing the heat seal marks, and dog hairs found on Helen's clothes were identical to hairs from Simms's Rottweiler and black Labrador retriever. Carpet fibers from the floor of Simms's apartment were also present on Helen's clothing.

It should have been enough to convince the most skeptical of juries, but given the absence of a body, investigators left nothing to chance. In the first example of its kind, they took blood samples from Helen McCourt's parents in hopes that DNA profiling would match them to blood found at the pub. All three codes matched. After some complex calculations, Dr. Alec Jeffreys declared that the blood from the pub was 14,500 times more likely to be from a child of the McCourts than a random sample from the population.

On March 14, 1989, Simms was jailed for life.

Conclusion

Tragically, the whereabouts of Helen McCourt's body remains a secret known only to Simms. As Dr. Moore reflected, "The only way the body will be found now is if he tells us . . . or it turns up by accident."

EXPLOSIVES AND FIRE

Bombs and fire have always held great appeal for the criminal, presumably in the belief that the secrets of their activities will be consumed in the conflagration. Nothing could be further from the truth. The killer, anxious to dispose of the victim, will find that a human corpse is extraordinarily fire resistant, requiring extreme temperatures to destroy teeth and bone. The bomber, too, is equally handicapped. If the crime scene has been processed correctly, modern laboratory techniques can isolate the most arcane substances. The accomplished technician knows exactly what items and substances will be needed for analysis. Highest on the list are any undetonated samples of explosive: These are most likely to be found close to the source of the blast. For this reason, all of the soft and porous material from this area, such as wood, soil, clothing, and furniture, is collected in case it has been penetrated by explosive residues.

Once in the laboratory, this debris is first microscopically examined to categorize the different types of evidence and to search for unconsumed traces of the explosive. Once the useful material has been removed, the debris is cleaned with acetone, which absorbs most explosives and can be used for chemical testing. The most common technique for analyzing and identifying explosives is chromatography.

In gas chromatography, which is used for analyzing fire residue, a sample is vaporized and sent down a tube by an inert gas, where all the compounds in the sample separate. Because accelerants and explosive chemicals move down the tube at different rates than do other residue substances—usually ash, charred wood, plastic, or fiber—each compound settles in the tube at a different rate, called the retention time. By consulting a reference table of retention times, the analyst can identify each compound.

For compounds that would break down completely if vaporized, liquid chromatography, in which the sample is carried down the tube by a liquid, is used. Thin-layer chromatography also uses a liquid to transport the sample. In this technique a glass slide is coated on one side with a thin layer of finely ground silica gel, and spots of the sample are placed along the bottom part of the slide. The slide is then partially immersed in a liquid, which by capillary action climbs up the slide and over the sample, separating its components. At this point, the separations are generally colorless and must be either viewed under ultraviolet light or sprayed with a suitable reagent, which will colorize the result. The resulting chromatographic plate is then compared with other samples or a control substance for similarities.

Despite the impressive array of analytical techniques that can identify not only the components but the origin of an explosive device, bombs remain inexpensive to construct, easy to conceal, and virtually impossible to guard against. They can slaughter thousands and profoundly affect the lives of millions more. Whether domestic or international, the dedicated terrorist has easy access to a frightening array of explosive devices. The fertilizer-based bomb that worked so dreadfully in Oklahoma in 1995 was similar in construction to the device used at the World Trade Center two years earlier. On that occasion the twin towers withstood the assault, but as the ghastly events of September 11, 2001, demonstrate, nowadays the suicide bomber doesn't even have to construct his own device; others do it for him. Twenty thousand gallons of high-octane jet fuel encased in pressured containers, traveling at several hundred miles per hour, need only the intervention of a deranged hand to become the deadliest criminal use of explosives yet seen.

Frederick Small

DATE: 1916

LOCATION: Mountainview, New Hampshire

SIGNIFICANCE: An almost foolproof plan to commit the perfect murder failed because of the desire to save a few dollars.

At fifty, the impression that Frederick Small liked to convey was that of a successful retired broker who had brought his considerable fortune to the shores of Lake Ossipee in Mountainview, New Hampshire, where he lived in connubial bliss with his devoted wife, Florence. Sadly, the truth fell short of this rosy scenario: All of Small's business ventures had failed ignominiously, he was perennially broke, and Florence was the third woman to bitterly regret walking down the aisle with this sadistic and greedy psychopath.

Small spent most of his time tinkering in a workshop at the rear of his cottage. Newfangled gadgetry fascinated him, and he was constantly trying to invent some product that would make his fortune. A conspicuous lack of success in this area meant that he had to keep a very watchful eye on the meager investments he did have, and every few weeks or so he would make the hundred-mile train journey to Boston to check on his fast-diminishing portfolio.

September 28, 1916, was such a day. In the morning he phoned the local coachman, George Kennett, and asked to be picked up that afternoon and taken to the railroad station. At 3:30 P.M., as arranged, Kennett collected Small's mail from the rural delivery box and then proceeded eagerly to the cottage. It was a stop he always looked forward to; invariably, before setting off, Small would invite him in for a tot of rye. Today, though, Small was waiting outside. Hardly glancing at the letters, he tossed them indoors, called out "Good-bye" to Florence, and slammed the door shut. Then the two men drove off.

At the station, Small was joined by Edwin Conner, a local schoolteacher and insurance agent who had agreed at late notice to accompany him to Boston. Together they caught the four o'clock train. After checking into a Boston hotel, Small mailed his wife a postcard at 8:40 P.M.—he wrote the time on the card—and then the two men took in a movie. They returned to the hotel around midnight, at which time the desk clerk informed Small that he should phone Mountainview at once—a fire had been reported at his lakeside cottage. Over the phone,

Small learned that his house was still ablaze and that Florence was believed to be inside. He limply handed the phone to Conner, who heard the tragic message repeated.

Both men raced back to Mountainview, driving through the night in a hired car. At dawn, distraught and inconsolable, Small viewed the smoldering ashes of his cottage, mumbling that it must have been the work of a passing itinerant. When it was safe enough to do so, a search party entered the blackened shell. In the cellar, they found what was left of Florence Small. The arms and legs were partially destroyed by flames, but the head and trunk were intact. When informed of the discovery, Small gasped, "You mean there's enough to be buried?"

Savage Murder

Indeed there was. And there was also enough to show that Florence had been murdered well before being engulfed in flames. The cord knotted tightly about her neck may have killed her, as might the .32 caliber bullet found in her skull, but the murderer had made triply sure by bludgeoning Florence Small's face to an almost unrecognizable pulp. Whoever burned the cottage obviously intended to destroy all traces of the crime and would have succeeded had not the exceptionally fierce blaze incinerated the floorboards, plunging Florence's body from the bedroom into the basement and several inches of water, where it was shielded from the searing heat.

An explanation for the fire's unusual intensity—it even melted the basement stove—came when George B. Malgrath, the medical examiner, noticed a strange resin on the body. It turned out to be the slag left by thermite, a welding compound of aluminum filings, metallic oxides, and magnesium, that when ignited produces temperatures in excess of five thousand degrees Fahrenheit. In seeking to annihilate the body, the killer had clearly overplayed his or her hand.

When officers searching among the rubble found a .32 Colt pistol that belonged to the bereaved husband, suspicion turned sharply in Small's direction. But, he protested, he had been one hundred miles away when the fire had started at ten o'clock.

Already, though, something else had turned up among the wreckage—the twisted remains of an alarm clock with wires and spark plugs attached. Also, police learned that just hours before Small's Boston trip, five gallons of kerosene had been delivered to him. Given these components and his technical knowledge, the construction of a device timed to explode when he was in Boston would not have been difficult. Considerable support for this theory came when Dr. E. W. Hodgson found the contents of Mrs. Small's midday meal still in her stomach, indicating that she had died hours before the fire started.

Violent Marriage

As always, the question of motive figured prominently in official considerations of whether to charge Small. His brutality toward Florence was well documented—on one occasion, he had to be restrained from beating her with a boat oar—and neighbors had become used to the sound of her screams echoing across the lake at night. But in the end, it was Small's old nemesis, money, that trapped him. Less than a year earlier, he had taken out a twenty-thousand-dollar joint insurance policy on their lives. He had originally attempted to do this without Florence's knowledge, until the agent, Edwin Conner, pointed out the impropriety of such an arrangement. Brought into the discussions, the hapless woman had willingly signed the policy. At the same time, Small insured the cottage for a further three thousand dollars.

Small continued to protest his innocence, playing what he considered to be his trump card: his insistence that Kennett, the coachman, had seen Florence on the veranda as they left for Boston. It was a shrewd psychological ploy, and it might have succeeded with someone more impressionable. But Kennett was adamant: No, he had not seen Florence, he had only heard Small's shouted good-bye. Small was finished.

During Small's trial at Carroll County Courthouse, prosecutors presented a mechanic who contrived an arson machine on the spot from apparatus identical to that salvaged from the fire. On January 8, 1917, Small was found guilty and sentenced to death. One year later, he paid the ultimate price on the gallows at the State Prison in Concord.

Conclusion

Ironically, Small's downfall was brought about by his own miserliness. Some time earlier, after a dispute with a contractor, he had rebuilt a section of the cellar ceiling himself, using cheap boards, and it was through this very section that the bed supporting his wife had tumbled. Had the floor held, then the body would have been entirely consumed by flames and Frederick Small would have been free to spend his insurance money, content in the knowledge that he had committed the perfect murder.

Charles Schwartz

DATE: 1925

LOCATION: Walnut Creek, California

SIGNIFICANCE: This case presents a vivid example of Edward O. Heinrich's ability to tackle every aspect of criminal investigation.

The small company founded by Charles Schwartz in Walnut Creek, California, was run with but one thought in mind—to invent an artificial silk that was indistinguishable from the genuine article. Such dreams did not come cheaply, and by the summer of 1925, Schwartz had sunk considerable sums of money, mostly his wife's, into the venture. Her belief in him didn't waver, and vindication came when one day he emerged from the laboratory, triumphant. The samples he held aloft were breathtaking. All who held the fabric compared it favorably to Chinese silk. During the celebrations, though, Schwartz injected a note of caution, hinting that certain unnamed businessmen would stop at nothing to get their hands on his invention. He intended to remain vigilant.

Anyone dismissing this as entrepreneurial paranoia received the rudest of shocks on the night of July 30, 1925, when the laboratory was shaken by a tremendous explosion, then burst into flames. As firemen dragged the charred body of a man from the debris, it was impossible to forget Schwartz's fearful warning. His wife, with only some jewelry and a watch to guide her, identified what was left of the corpse. The presence of several incendiary devices made it clear that this had been no accidental blaze, especially when a neighbor reported seeing a car similar to Schwartz's race away after the explosion. An unanticipated complication arose with the discovery that just before the fire, Schwartz had insured his own life for $185,000, a vast amount in the 1920s. Understandably, the insurance company wanted confirmation that the dead man was actually Schwartz.

Given his proximity to the crime and the assistance he had rendered in the 1923 Southern Pacific Railroad murders (see the D'Autremont Brothers case on page 299), it was only natural that University of California scientist Dr. Edward O. Heinrich would be called in to assist the inquiry. He asked for the charred body to be sent to his laboratory.

Even before Heinrich became involved, events had taken another peculiar twist. Mrs. Schwartz told police that one day after the explosion, someone had burgled her house and stolen every photograph of her husband. There had been no attempt to take anything else. Heinrich, who had requested just such a photograph to aid in identification,

Edward O. Heinrich, America's finest early forensic scientist, is credited with solving more than two thousand crimes.

asked if any other photographs of Charles Schwartz existed. After a moment's thought, Mrs. Schwartz recalled that her husband had recently had his portrait taken in Oakland. The photographer there found the negative and rushed an enlargement to Heinrich.

Heinrich already knew that he was dealing with a case of murder—the skull had been battered repeatedly. Now, one glance at the photograph convinced him that the corpse was not that of Charles Schwartz. In the photograph, Schwartz had an instantly recognizable mole on one earlobe. This corpse did not. There were other discrepancies. According to his widow, Schwartz had eaten a meal of cucumbers and beans just before the explosion. Yet when Heinrich examined the stomach contents, all he found was undigested meat. And microscopic examination of hair from Schwartz's hairbrush revealed clear differences from that of the dead man.

Switched Body

The murderer had gone to considerable lengths to create the impression that the body was that of Charles Schwartz. A molar missing from the right upper jaw matched Schwartz's old dental chart, until Heinrich discovered that this socket was still raw, indicating a tooth pulled shortly before or after death. Significantly, other identifying features were obliterated: Someone had gouged out the eyes, and the fingerprints had been destroyed with acid.

Scraps of paper found in a closet at the burned-out laboratory were put under a microscope. Just visible was the name "G. W. Barbe" and a partial address, "Amarillo, Texas." Gilbert Warren Barbe was an itinerant preacher who had drifted into town and become friendly with Schwartz. Curiously, the two men were very similar in appearance, and now Barbe had disappeared. When Heinrich learned from Mrs. Schwartz that her husband's eyes had been brown, while those of Barbe were blue, he was convinced that the body in the fire belonged to Barbe and that Schwartz was the man seen fleeing in the car.

Heinrich suspected that Schwartz had been planning this crime for months, just waiting for the right victim to appear. He doubted that the so-called lab had ever been used for that purpose; it lacked water and gas, and the bills showed only enough electricity consumption to keep a single light burning. It had all been a sham.

Mrs. Schwartz would hear none of it, angrily brandishing the artificial silk as proof that her husband had been well intentioned. Sadly, the police had to inform the distraught woman that her husband's supposedly artificial silk was, in fact, genuine and had been bought in a San Francisco store. They also showed her an advertisement that Schwartz had placed for a chemist's assistant. Among the qualifications requested were small hands and feet! Clearly, Schwartz had been advertising for a victim. And then he had met the traveling missionary, Warren Barbe, a man remarkably similar to himself and someone unlikely to be missed. Schwartz had gambled that their three-inch difference in height would pass unnoticed, because fire often shrinks its victims, but he was wrong. When Heinrich measured the corpse, it was still three inches longer than Schwartz had been in life.

Although the police were certain that Mrs. Schwartz had not been privy to her husband's wickedness, they also realized that Schwartz's only hope of getting the $185,000 would be to contact her when the excitement had died down. To this end, they encouraged local newspapers to report that the claim had been settled. The story received wide coverage and caught the eye of an Oakland rooming house landlord who thought that the deceased man bore a striking resemblance to one of his boarders, a Mr. Harold Warren, who had rented a room one day after the explosion.

A call to the police brought a swift response. But when one of them rapped on the stranger's door, the only answer was a single shot. They burst into the room to find Charles Schwartz dead on the floor, a smoking revolver in his hand. On a table was a note addressed to his wife; in it he confessed his guilt and begged her forgiveness.

Conclusion

Edward Heinrich's grasp of so many varied disciplines makes him a unique figure in the history of forensic science. He was a geologist, a physicist, a handwriting expert, an authority on inks and papers, and a biochemist. But perhaps his greatest strength lay in deductive reasoning—a gift so strong that he became known as the American Sherlock Holmes.

Sidney Fox

DATE: 1929

LOCATION: Margate, England

SIGNIFICANCE: The arguments fostered by this cause célèbre prompted a courtroom conflict that is discussed to the present day.

When sixty-three-year-old Rosaline Fox and her illegitimate son, Sidney, thirty, checked into the Metropole Hotel in Margate on October 16, 1929, they were penniless and without a scrap of luggage. This was not an uncommon state of affairs for this most uncommon of couples. Behind them lay a decade-long trail of bad checks, unpaid bills, angry hoteliers, and frequent brushes with the law. In between jail sentences, Sidney Fox fancied himself something of an English upper-class gentleman. He was plausible and cunning and proudly homosexual, though pragmatic enough to adopt more conventional forms of seduction should the flagging family coffers so demand. In these endeavors, he enjoyed his mother's unstinting support.

On October 22, he scrounged the rail fare to London and called at an insurance office to arrange for two accident policies on his mother's life—with an aggregate value of three thousand pounds (twelve thousand dollars)—to be extended until midnight the following day. Then he returned to the Metropole Hotel at Margate.

The next evening, at about 11:35 P.M., fire broke out in the adjoining rooms occupied by Fox and his mother. Fox, clad only in a shirt, ran along the corridor, yelling for help. He directed another guest toward his mother's room. Clutching a handkerchief to his face, Samuel Hopkins bravely entered the smoke-filled room, past a furiously burning armchair, to the bed where Mrs. Fox lay unconscious. He dragged her outside. Fox, who made no attempt to assist, watched from a safe distance. Two doctors examined Rosaline Fox and pronounced her dead from shock and suffocation. Sidney, it is reasonable to assume, must have glanced at the clock: His mother had died just twenty minutes before the insurance coverage was due to expire.

A few days later, Mrs. Fox was buried in her native East Anglia. Fox went directly from the graveyard to the insurance company, where he filed his claim with such an absence of funereal solemnity that a suspicious agent telegraphed his head office to suggest that they investigate this case more closely.

Subsequently, the matter was referred to Scotland Yard and an exhumation order obtained. The postmortem was performed by Sir Bernard Spilsbury, by then a national figure at the peak of his celebrity. The absence of carbon monoxide in the blood or sooty deposits in the air passages convinced him that Mrs. Fox had died before the fire started, and he believed that she had been manually strangled. His reason for thinking this was based solely on one factor—the alleged presence of a bruise at the back of the larynx. There were none of the usual physical signs associated with strangulation—no finger marks on the neck, no petechiae (pinprick bruising of the face), no fracture of the brittle bone in the throat known as the hyoid—yet Spilsbury was adamant that Mrs. Fox had been throttled.

Second Opinion

Fox was duly arrested and charged with murder. At the request of the defense, another eminent pathologist, Professor Sydney Smith, examined sections from the larynx that Spilsbury had preserved in formaldehyde. Apart from some putrefaction, Smith couldn't find anything at all, certainly not a bruise, and told Spilsbury so. Unaccustomed to having his opinion doubted, the older man snapped, "It must have faded . . . it was there when I made my examination."

Under cross-examination Spilsbury became even crustier, refusing to countenance the possibility of a mistake. "It was a bruise and nothing else. . . . There are no two opinions about it!"

Professor Smith, who believed that death was due not to strangulation but to heart failure precipitated by the smoke and fire, testified for the defense. Attorney General Sir William Jowitt wasted no time in raising the subject of the vanishing bruise: "Sir Bernard says there can be no two opinions about it."

"It is very obvious that there can," replied Smith silkily.

"But you are bound to accept the evidence of the man who saw the bruise?"

"I do not think so."

"How can you say that there was no bruise there?"

"Because if it was there, it should be there now. It should be there forever."

It was unanswerable logic, but Spilsbury's reputation by that time had reached near-mythic proportions. To the British population at large, from whom all juries were drawn, his word had an almost biblical authority. The dangers inherent in such a situation were admirably pointed out by defense counsel J. D. Cassells, QC, who said, "It will be a sorry day for the administration of justice in this land if we are to be thrust into such a position that because Sir Bernard Spilsbury expressed an opinion, it is of such weight that it is impossible to question it."

Source of Blaze

Far more damaging, from a defense point of view, was the mystery of just how the blaze had started. Dr. Robert Nichol, who had attended Mrs. Fox, noticed that "there seemed to be no connection between the obvious source of the fire and the site of the fire itself." He was referring to the fact that flames had seemingly leaped from the gas fire to the armchair without burning any of the carpet in between. Margate's chief fire officer expressed his opinion that "the fire unquestionably originated directly underneath the armchair." His attempts to replicate the blaze had necessitated the use of gasoline, because the carpet would not burn otherwise. Coincidentally, Fox had had a bottle of gasoline in his room, which had been used, he said, to clean his only suit.

If ever any defendant should have availed himself of the right to silence, it was Sidney Fox. But that was not his way. His testimony, shiftily given and teeming with the most transparent lies, proved deadly. Asked to explain why he had slammed shut the door of the burning room where his mother was fighting for her life, he replied feebly, "So that the smoke would not escape into the hotel." A wave of revulsion swept the court as they stared at someone prepared to jeopardize his own mother's life in such a manner.

In his final charge to the jury, the judge, although clearly repulsed by Fox's performance, expressed grave concerns about Spilsbury's testimony. "There were no marks on the throat. Sir Bernard Spilsbury had said it was quite possible there would be none, but you and I might find that hard to believe. As regards the hyoid, it is a very curious coincidence that it was not broken in this case. That is a very strong point in favor of the accused."

Listening to these words, Fox may have retained some hope. If so, it was sorely misplaced. After just an hour's deliberation, the jury pronounced him guilty and he was hanged at Maidstone on April 8, 1930.

Conclusion

So was Fox guilty of murder? All the available evidence seems to suggest that he had intended to arrange an accident for his mother, that is, to burn her to death. In light of this, Spilsbury's verdict of strangulation must have come as a bombshell to him. Ironically, Spilsbury's intransigence offered the defense their only hope of securing an acquittal, and had Fox not testified and exposed himself in so callous a fashion, then the bitter forensic wrangling might well have led to an acquittal.

John Graham

DATE: 1955

LOCATION: Denver, Colorado

SIGNIFICANCE: Painstaking reconstruction of the crime allowed scientists to solve the worst single incident of aeronautical mass murder on American soil.

Just eleven minutes after taking off from Denver on November 1, 1955, United Airlines flight 629 to Portland, Oregon, exploded, killing all forty-four people on board and cascading debris over miles of countryside. A team of investigators from United Airlines, the Douglas Aircraft Company, and the Civil Aeronautics Board was charged with discovering whether the crash came about because of mechanical failure, human error, or sabotage.

They began by superimposing a vast grid of squares over the entire crash scene. Searchers combed this area for scraps of metal, luggage—anything that came from the stricken plane. Every item and its location was noted on the grid before it was shipped to a Denver warehouse, where it was wired to a wooden mock-up of the DC-6B. As the fuselage took shape, it became obvious that no pieces could be found to fit a jagged hole near the starboard number four cargo bay.

The metal at this point—bent outward by a force far more powerful than just a crash—showed ominous signs of scorching; similarly discolored shards had been driven upward through the soles and heels of the passengers' shoes. Clearly, a violent explosion had occurred. Because there were no fuel lines or tanks in this part of the plane, the blast had to have been caused by either an accidental cargo explosion or deliberate sabotage.

On November 7, the FBI was asked to initiate a criminal investigation. Within twenty-four hours, some one hundred agents were deployed, either digging into the backgrounds of the plane's crew and passengers or checking air freight manifests to determine whether there had been any illegal shipments of potentially explosive materials. This latter line of inquiry drew a complete blank, but a glimmer of light began to illuminate the path ahead.

Airline records showed that a passenger, Mrs. Daisie King, en route to visit her daughter in Alaska, had checked a suitcase, but apart from a few barely identifiable fragments of leather, this case had seemingly vaporized in the crash—unlike the bag Mrs. King had carried on board as hand luggage. Tucked into one of the pockets of that bag, agents found a four-year-old newspaper clipping about someone called Jack Graham, whom authorities were seeking due to allegations of forgery. Investigators soon learned that Graham was Mrs. King's twenty-three-year-old son. As well as the charge of forgery, he had also been arrested for theft, but the charges were dropped after his wealthy mother had repaid the missing money. Since getting married, the wayward youth appeared to have mended his ways. Even so, investigators couldn't overlook the fact that Graham had chauffeured his mother to the airport on the fateful day of the plane crash.

Like all of the victims' relatives, Graham was interviewed. He spoke to agents first on November 10. Questioned about the contents of his mother's luggage, he said that she always insisted on packing everything herself and would never let anyone help her. As an afterthought, he did mention that she had packed some shotgun shells and rifle ammunition for a hunting trip in Alaska. When asked specifically if he had put anything into his mother's luggage, Graham replied that he had not.

Suspicious Gift

Graham's wife, Gloria, verified Mrs. King's single-mindedness about packing her own luggage but added something else. Just minutes before setting off for the airport, her husband had taken a bulky gift-wrapped package to the basement. The present, she thought, was a jewelry tool kit intended for his mother to use in her hobby of fashioning trinkets from seashells.

This was news to the agents; Graham had said nothing about any gift. But a neighbor had also heard Graham mention the tool kit—how he had searched all over for just the right kind, then gift-wrapped it and placed it in his mother's luggage as a surprise for her when she reached Alaska. Inquiries made in the Denver area found only two stores that stocked kits suitable for cutting seashells, and neither reported selling such an item during October.

At a second interview, on November 13, Graham denied knowledge of any gift for his mother. That same afternoon, FBI scientists reported that burned metal retrieved from the crash scene showed traces of sodium carbonate, sodium nitrate, and sulfur-bearing compounds—the residue left by an explosion of dynamite.

Agents quickly obtained a warrant to search Graham's house, where they found not only the shotgun shells and ammunition that Mrs. King had supposedly packed in her luggage, but Gloria Graham, confused and very frightened. Jack, she sobbed, had ordered her to say that she had been mistaken about the present.

Faced with these revelations, Graham suddenly recalled purchasing the tool set for his mother. He had paid a stranger ten dollars for it in a transaction witnessed by two employees at the garage where he worked. After slipping the gift-wrapped package, bound with transparent tape, into his mother's suitcase, he had stashed the remainder of the tape in his car.

While Graham blustered, FBI agents ransacked his house. In one of his shirt pockets they found a small roll of copper wire of a type used for detonating explosives. Even more damning, hidden in a cedar trunk was a $37,500 life insurance policy purchased by Daisie King at the airport just minutes before takeoff and made payable to Jack. Nobody had suspected this. Earlier, insurance companies had been asked for details of all the life coverage bought by passengers on flight 629; somehow Mrs. King's policy had been overlooked.

The incongruities kept surfacing: Investigators found some stockings and a cosmetics bag belonging to Mrs. King and gifts that she had intended for her daughter in Alaska. Surely she would have taken these things with her. When asked why his mother had left them behind, Graham mumbled weakly, "I told her not to take them because her baggage was overweight."

Cool Confession

At midnight, news came that Graham's two co-workers had no recollection of any tool kit purchase. Shortly afterward, for the first time, Graham was told of the FBI report stating that the crash was caused by a dynamite explosion. Graham digested this information for a moment while sipping from a glass of water. Finally, he shrugged: "Okay, where do you want me to start?"

Without a flicker of emotion, he told of constructing the bomb with twenty-five sticks of dynamite, two electric primer caps, a timer, and a six-volt battery. In order to acquire the necessary bomb-making expertise, he had worked at an electronics shop for ten days at $1.50 an hour. As suspected, he had taken several items from his hapless mother's suitcase and replaced them with the lethal device.

From the moment of his arrest, Graham's only concern was to impress upon everyone just how much he loathed the mother who had abandoned him in his childhood and then reclaimed him only when she had remarried much later. But the financial angle could not be disregarded: Besides the insurance policy, Graham also stood to benefit from his mother's $150,000 estate.

At his trial, which was broadcast on TV, Graham recanted his confession and pleaded not guilty. Throughout, he remained impassive and unremorseful. Verdict and sentence were foregone conclusions, and on January 11, 1957, without saying a word, Graham walked to the gas chamber and died.

Conclusion

Had Graham been a student of criminology, he might well have forsaken aero-explosions as a means of killing his mother. Just six years earlier, a Canadian named Albert Guay had employed identical tactics to rid himself of an unwanted, overinsured spouse, with equally calamitous results for everyone concerned. Twenty-three lives were sacrificed so that Guay and two accomplices might pocket a few thousand dollars. They didn't get to spend one cent; all three were hanged.

Steven Benson

DATE: 1985

LOCATION: Naples, Florida

SIGNIFICANCE: Even an electronics genius found that he was no match for modern forensic techniques in this sensational saga of familial feuding that ended in tragedy.

At quarter to eight on the morning of July 9, 1985, Steven Benson, a thirty-three-year-old electronics whiz and heir to the family tobacco fortune, arrived at his grandmother's palatial home in the Quail Creek district of Naples, Florida. Together with his relatives, he intended to stake out a lot for a new house. After loading stakes and blueprints into the family's 1978 Chevrolet Suburban, he drove to a local convenience store, where he bought coffee and sweet rolls for the other fam-

Charred wreck of the Chevrolet Suburban after the Benson family murders.

ily members. Just before nine o'clock, Steven accompanied his mother, Margaret Benson; his sister, Carol Lynn Kendall; and his adoptive brother, Scott, out to the Suburban. As they were about to drive off, Steven suddenly remembered that he had left a tape measure in the house. Tossing the keys to Scott, he headed indoors. Seconds later, Scott turned the key in the ignition and a massive orange flash enveloped the vehicle.

The Suburban blew apart. Only Carol Lynn survived the blast. As she crawled away from the twisted shell, helped by golfers from the adjacent golf course, a second blast erupted, turning the Suburban into an inferno.

The murder of such socially prominent citizens—the Bensons were very rich—sent shock waves throughout Naples, and by 11:30 A.M., agents from the Bureau of Alcohol, Tobacco, and Firearms (ATF) were on the scene. Amid the twisted wreckage, scorched wire, and splinters from the wooden stakes, they found obvious signs of a crude bomb: fragments of galvanized pipe and two end caps. An interested observer of all this activity, Steven Benson seemed remarkably nonchalant for someone who had just had a close brush with almost certain death.

When ATF explosives expert Albert Gleason began his examination of the shattered vehicle, he soon noted the sources of the two explosions: one in the console area, the second to the rear. This supported the conclusions of Collier County Medical Examiner Heinrich Schmid, who found that Margaret Benson's injuries were mostly on the left side of her body and Scott's were on the right, indicating that at least one bomb had exploded between them.

A little-known feature of explosions is just how much of the actual bomb is left intact after a blast. On this occasion, Gleason established that the two bombs were each about a foot long, with four-inch end caps screwed onto pipes to seal the explosive. Two of the end caps were stamped with the manufacturer's mark, one with a letter *U* for Union Brand, another with *G* for Grinnell, raising the possibility that both seller and buyer might be traced.

Forensic Dig

Gleason returned to the bomb site and asked that an area twenty-five feet in diameter be excavated around it. In the manner of an archaeological dig, officers first sifted dirt through a wire strainer and then pawed at it with magnetized gloves. Pipe bombs are simple devices; their primary components are a canister (in this case a galvanized pipe threaded at both ends), an explosive device, and some kind of detonator. Gleason had already identified the bomb type; now he wanted to know how it had been triggered.

Close to the explosion, the remains of four 1.5-volt D-cell batteries were found. Gleason estimated that an electronically controlled switching device powered by six volts would detonate the bomb. Eventually, among the debris, he found a small piece of a circuit board that did not appear to belong in the vehicle, and there was also a manual switch that couldn't be accounted for.

Meanwhile, other investigators visited all the local hardware stores, construction sites, plumbing supply shops, and junkyards, asking about end caps and sections of galvanized pipe. One Naples store checked its records and found that it had sold two Union Brand end caps on July 5, and two foot-long sections of threaded pipe on July 8. The sales assistant described the customer as about six feet tall and two hundred pounds, very similar to Steven Benson.

Fingerprint analysis of the pipe bomb components came up blank. Laser light tests, too, failed to produce anything of value. But then someone suggested checking the two sales tickets at the supply store. These were treated with chemicals. On the face of each was a latent palm print, what experts term a "writer's palm," made by someone who was left-handed, someone like Steven Benson. After considerable legal wrangling, agents managed to obtain inked impressions of Benson's fingerprints, which matched those on the tickets.

Career Fraudster

A background check revealed that Benson had been defrauding his mother for years. Recently his dishonesty had come to light, leading to threats of disinheritance. Before he would let that happen, agents speculated, Steven had decided to eliminate his family so that he might inherit the entire fortune, estimated at ten million dollars. With his electronics expertise, the construction of a pipe bomb would be simplicity itself, as would the construction of a detonator that he could trigger when he was a safe distance from the car. On August 21, after much legal consideration, Benson was arrested.

Eleven months later, he stood trial for murder. Although Carol Lynn remembered that the blast had followed almost immediately after Scott turned the key, Gleason expressed doubts that the ignition switch had triggered the explosions. Although he admitted that he was never able to locate the detonator, those scraps of scorched circuit board and batteries made it virtually certain that the blast had been detonated from outside the car. To give some idea of just how powerful the bomb had been, Gleason presented what he believed to be a similar device. Empty, the length of pipe weighed twenty pounds, and could have held anywhere between three to six pounds of high explosive.

With the fingerprint evidence as well, it all made for a damning case. On August 7, 1986, Steven Benson was found guilty of first-degree murder and sentenced to life imprisonment.

Conclusion

Bombs are notoriously resilient to the effects of their own destructiveness. Invariably they leave clues, and as Steven Benson discovered, where there are clues, there is usually a path to the bomb maker.

Pan Am Flight 103

DATE: 1988

LOCATION: Lockerbie, Scotland

SIGNIFICANCE: Spanning three continents, the forensic inquiry spawned by this terrorist massacre was, at the time, the most exhaustive ever mounted.

On December 21, 1988, Pan Am flight 103 departed London's Heathrow Airport en route to New York. The passengers, mainly American students and servicemen on their way home to celebrate Christmas, were in good spirits. As the jet winged northward, it climbed steadily. At thirty-one thousand feet, catastrophe struck: An explosion blew the Boeing 747 apart. Six miles below, the tiny Scottish hamlet of Lockerbie lay helpless. When flight 103 slammed into the ground, a huge fireball erupted, killing all on board and numerous townspeople, a total of 270 lives.

There had been no radio contact from the plane; one second it was on the radar screens, and then it had disappeared. Within hours, investigators from the FBI, Britain's Air Accidents Investigation Branch (AAIB), and local police were scouring the scene for clues and survivors. Only in the full light of day was the extent of the tragedy made apparent; debris from the plane had littered 845 square miles of the Scottish Borders and northern England. It was soon evident that no structural defect could have caused such widespread devastation; it had to have been a bomb.

Early the next morning two children found a box with a metal handle. It measured two feet by six inches, and on its side was printed "DATA REPRODUCER 1972." This was the plane's "black box," or flight information recorder; from it, experts hoped to unravel the mystery of what had brought down flight 103.

To aid them in the gruesome task of identifying victims, investigators enlisted the aid of the Home Office Large Major Enquiry System (HOLMES), a sophisticated computer program designed for the British police, capable of handling vast amounts of data. Every scrap of information was fed into the program: location of a body, type and nature of injury, clothing color and style. All were logged, sorted, and cross-referenced. For the FBI team it meant an opportunity to test the effectiveness of their latest forensic weapon: a phototelesis machine that could transmit and receive crystal-clear images of fingerprints and other identifying body marks over commercial phone lines, much like a fax machine but much sharper.

Top Laboratory

Christmas Day saw the discovery of an oddly misshapen piece of metal. It was sent to the Royal Armament Research and Development Establishment at Fort Halstead in Kent, in all probability the most sophisticated forensic crime laboratory in the world. Equipped to the same standards as NASA, Fort Halstead's workbenches have zero static electricity, which allows for virtually perfect test conditions. Scientists there identified the twisted piece of metal as part of a baggage container, and they were unanimous in their conclusion—only a bomb could have caused such warping. While the faint possibility of a lone assassin operating from personal or financial motives could never be entirely ruled out, in all probability Pan Am 103 had been a terrorist target.

Intelligence services on both sides of the Atlantic had suspected as much from the outset, but trying to fix responsibility among the ranks of the desperate and the disaffected was no easy matter. More data was needed, and for that the scientists and crash investigators would have to establish the bomb's type and likely point of origin.

In a vast hangar, technicians began reassembling the 747 piece by piece, a forensic jigsaw puzzle of unparalleled complexity. Ultimately, 85 percent of the plane was rebuilt. As in all such reconstructions, the negative evidence, or the pieces that were missing, offered the strongest lead. A fifty-foot section of fuselage, housing the forward luggage hold, could not be found: Any explosion in this vicinity would have devastated the plane's main electrical generator, destroying all hope of radio contact. After hours of painstaking work, investigators were able to pinpoint the blast to cargo bay 14L.

From among the thousands of fragments, scientists recovered three minuscule pieces of a gray radio–cassette player, similar to that used by terrorists in West Germany as a means of bomb concealment. They also found traces of the plastic explosive Semtex on what was left of a Samsonite copper-colored suitcase. All the indications were that the lethal radio–cassette player had been packed inside the suitcase, then detonated using an electrical circuit activated by a fall in atmospheric pressure.

Maltese Connection

Slowly the suitcase's origins came into focus. A scrap of blue fabric from a baby's jumpsuit had survived the crash, and on it was a label marked "Babygro." This was traced to a boutique in Malta, where the owner, Tony Gauci, recalled that several months earlier an Arab customer had purchased numerous items without regard to size or type. Because of the incident's peculiar nature, the owner had remembered it distinctly and was able to furnish a list of purchases to Fort Halstead. They included a tweed jacket, fragments of which were identified among the remains of the Samsonite suitcase, and an umbrella with fibers from the Babygro jumpsuit blasted into its fabric. Clearly, all these items had been in very close proximity to the bomb.

How the bomb actually got onto flight 103 remains a mystery. Computer records from a German company that operates the sophisticated international luggage transfer system suggest that a suitcase was loaded onto a flight from Malta to Frankfurt, Germany, and there placed on board flight 103. Who ordered the outrage is also an international guessing game. The name most often mentioned is Ahmed Jibril, head of the terrorist group Popular Front. One theory claims that the bomb was placed in retaliation for the downing of an Iranian passenger jet by the USS *Vincennes* a year earlier. Others feel that the targets were four American CIA operatives on board, men with considerable experience in Middle East hostage negotiations. Even more uncertain is how the deadly suitcase managed to evade security checks at both Frankfurt and London.

After a three-year joint investigation by Scottish police and the FBI, in which more than 15,000 witness statements were taken, indictments for murder were returned on November 13, 1991, against two Libyan intelligence officers, Lamen Khalisa Fhimah and Abdul Basset Ali al-Megrahai. The latter was identified by the Maltese storekeeper as the man who made those strange purchases just weeks before the Lockerbie disaster.

Initially, Libyan president Mohamar Ghadaffi resisted all requests for their extradition; only after several years of UN sanctions did he bend to international will. On April 5, 1999, the two former intelligence officers were handed over to Scottish police in the Netherlands, which had been chosen as a neutral venue. Their trial began on May 3, 2000, before a panel of three Scottish judges, acting without a jury. Libyan fears about the court's impartiality were allayed on January 31, 2002, when the three judges acquitted Fhimah. At the same time,

Megrahi was convicted of murder and sentenced to twenty-seven years in prison. After his appeal was refused, Megrahi applied to the European Court of Human Rights. They, too, decided not to intervene, and he continues to serve his sentence in a Scottish prison.

Conclusion

Prior to the events of September 11, 2001, bombs detonated by atmospheric pressure were a favorite terrorist device. In the wake of that tragedy, airport security systems worldwide tightened up their screening procedures, and baggage is now routinely subjected to low-pressure screening, in hopes of exploding any such concealed devices. This has been a significant factor in the longer check-in times that now affect airports everywhere. Passengers made impatient by this delay would do well to reflect on the sobering fact that some terrorist cells, in their zeal to circumvent these screening procedures, have been known to design circuits that delay the detonation of their deadly handiwork until it has been exposed to at least thirty minutes of low atmospheric pressure.

FINGERPRINTING

In 1879, Alphonse Bertillon, a clerk at the Prefecture of Police in Paris, produced the first systematic method of identifying individual criminals. Based on 243 separate body measurements, *Bertillonage,* as it became known, often proved surprisingly accurate, but it was doomed almost at birth by the simultaneous development of another means of identification, this one foolproof.

Nothing has more greatly enhanced the cause of crime detection than the discovery that no two people have the same fingerprints. This individuality had long been recognized in such civilizations as China and Babylon. A line in the book of Job reads, "He sealeth up the hand of every man that all men may know his work." But it was not until 1880, when a Scottish physician in Tokyo, Dr. Henry Faulds, published a letter in the British scientific journal *Nature* that modern fingerprint analysis began. Intrigued by finger impressions on fragments of ancient pottery, Faulds promulgated the idea that they might be a means of identification. The notion was roundly endorsed by Sir William Herschel, chief administrative officer of Bengal, who reported that he had used thumb impressions to identify illiterate prisoners and workers in India for years. The initial excitement sparked by the debate in *Nature* led to the first serious study of fingerprints by English scientist Sir Francis Galton.

After establishing that fingerprints were not inherited and that even identical twins had different ridge patterns, Galton laid the groundwork for a basic classification system, putting the most commonly observed features into three groups—arches, loops, and whorls. Galton's book *Finger Prints* appeared in 1892, but yet another Indian police officer, Sir Edward Henry, completed the task of fingerprint classification.

Henry established five basic patterns, adding tented arches to Galton's three groups and dividing loops into two classes. In 1896 he instructed the forces under his command to take up dactyloscopy (the term used for fingerprinting). Success with the new system was immediate; the rate of criminal identification in India soared dramatically.

Sir Edward Henry—the father of modern fingerprinting.

Fingerprint impressions fall into three basic types: latent, visible, and the plastic, or molded, print. By far the most frequent is the latent print, which is invisible to the eye. These are formed by sweat, either from the hands themselves or by unconscious contact between the fingers and the face or other parts of the body where the sebaceous glands are situated. Even if a criminal scrubs his hands and dries them thoroughly, if he then puts a hand to his face or his hair he is likely to leave a latent print, invisible to the naked eye, on anything he touches, particularly on such surfaces as glass or polished wood.

Treating, or developing, latent prints so they can be examined may be done in a variety of ways. The most common method is to use a gray or black powder (formerly a mixture of mercury and chalk, but nowadays organic to reduce the risk of health hazards to the user). Because they are more difficult to develop, prints on porous items such as paper and cardboard are best treated with either iodine fuming, which reacts with grease, or ninhydrin, which reacts with amino acids.

The second type of print, the legible kind, results from fingers stained with blood or ink or some other similar medium and is rarely found at the crime scene. The same goes for the third category, the plastic print, which is an impression made on a soft surface such as cheese, soap, or putty.

The durability of a latent print varies and is governed by several factors, but if it is made on a hard, protected surface and left untouched, it is virtually permanent. Latent prints have been found and developed from objects in ancient tombs.

Of course, a fingerprint is worthless without a reference database. This used to mean long hours of poring over card indexes. Computerization has changed all that. All the world's major law enforcement agencies maintain huge, ever-expanding fingerprint records, and what once may have taken weeks or even months is now accomplished in seconds.

For most of the twentieth century the sanctity of fingerprint identification remained unchallenged; it was the one item of evidence that police, courts, and juries agreed on. But recent developments on both sides of the Atlantic have raised troubling doubts. At issue lies one fundamental question: Is the identification of fingerprints—especially partial or smudged prints—a science or an art? Concerns arose in America when a 1995 proficiency test of

156 examiners conducted with the approval of the International Association of Identification, the profession's certifying organization, found that one in five examiners made at least one "false positive" identification—linking a mock crime scene print to the wrong person. In the real world, the 1997 Scottish murder trial of David Asbury was jeopardized by the insistence of a police officer, Shirley McKie, that she had not left a thumbprint at the crime scene (Asbury's defense hinged on the claim that the disputed print might belong to the real killer.) Despite McKie's denial, four experts from the Scottish Criminal Record Office (SCRO) testified that the print belonged to her. Because of their testimony, David Asbury was sentenced to life imprisonment and McKie faced charges of perjury. In 1999 McKie was acquitted after several international fingerprint experts declared that the original experts got it badly wrong: She had *not* made the print. This cleared the way for Asbury to appeal and, in August 2002, his conviction was quashed. In 2006, McKie was awarded $1.25 million in damages. Thus far, Asbury has not received a penny in compensation. The SCRO still refuses to acknowledge that its identification was flawed.

The McKie/Asbury fiasco has already been cited in several trials as an example of the fallibility of fingerprint identification. No objective assessment of the evidence could fail to condemn the SCRO's original blunder and its subsequent intransigence. The stakes are high. If official pigheadedness ends up sparking off a wholesale assault on what has always been the gold standard for human identification—the fact that in more than a century of usage, in every country on earth, no two people have ever been shown to share the same fingerprint—then the future of crime fighting looks bleak, indeed.

Francesca Rojas

DATE: 1892

LOCATION: Necochea, Argentina

SIGNIFICANCE: Criminals who have been convicted through the evidence of their own fingertips can trace their dubious heritage to this tragic, star-crossed girl who murdered for love.

At first glance, the town of Necochea, on Argentina's Atlantic coastline some 250 miles south of Buenos Aires, might appear to be an unlikely setting for criminological history. Yet it was here, in 1892, in a hut on the outskirts of town, that a double murder took place, the repercussions of which are felt to the present day. The victims were the two illegitimate children of a twenty-six-year-old woman named Francesca Rojas.

On the evening of June 19, Francesca burst into a neighbor's shack, screaming that Velasquez had killed her children. Velasquez, the neighbor knew, worked at a nearby ranch and had been an ardent if unsuccessful suitor of Francesca's. After sending his son for the police, the neighbor went to Francesca's hut and found the children, a boy of six and a girl of four, dead from massive head wounds. No weapon was found.

The chief of police spent little time examining the crime scene; he wanted to know more about Velasquez. Francesca sobbed that Velasquez, a middle-aged simpleton, was madly in love with her, but she had spurned his every advance. That afternoon the feebleminded ranch worker had called, more insistent than ever. Finally she had stunned him with the revelation that she had promised herself to another, younger man. Velasquez became apoplectic. Insults flew wildly. So did the threats of revenge. Then he rushed off. That evening, returning from work, Francesca found the front door wide open. As she approached, Velasquez had bolted from the hut, almost knocking her over in his haste to get away. In the bedroom, she found both children dead.

Taken into custody that night, Velasquez admitted having threatened Francesca but denied putting any such plans into practice. Even a severe beating could not budge Velasquez's claims of innocence. In frustration, the police chief opted for drastic measures. He had Velasquez bound and laid beside the illuminated corpses of the children all night in hopes that remorse and fear of retribution would loosen his tongue.

The next morning Velasquez was unshaken; neither did another week of relentless torture elicit any change in his story. Whether through weariness or a growing sense that he might be investigating the wrong man, the police chief began to look elsewhere. Already rumors were circulating about Francesca and her young lover, who had previously been overheard saying that he would marry her were it not for those two children.

While the police chief's suspicions were now focused elsewhere, his investigative techniques had lost none of their flair. He sought to jolt Francesca's nerves by spending an entire night outside her hut, rattling the windows and doors and crying out in what he imagined the voice of an avenging angel might sound like. But Francesca was made of sterner stuff than that. At dawn she left for work, her conscience intact, none the worse for the chief's nocturnal nonsense, and still she was adamant that Velasquez was the killer.

Call for Assistance

At this point, assistance was requested from the regional headquarters at La Plata. This was a stroke of immense good fortune, for overseeing the Bureau of Identification was a Dalmatian immigrant named Juan Vucetich, who earlier than most had seen the advantages of fingerprinting over Bertillonage. In September 1891, using Galton's basic material, he had formulated a ten-finger classification system. Thus far, its use had been restricted to inmates at the local prison, but Vucetich was eager to spread the gospel. When he dispatched Inspector Alvarez to investigate the murder at Necochea, though, it could hardly have crossed his mind that history was about to be made.

The level of police incompetence that Alvarez uncovered was astounding. Velasquez, he soon learned, had an unshakable alibi for the time of the murder, but the poor man, with his limited intelligence, had neglected to mention it. Francesca's young lover, too, was absolved from complicity. Which, Alvarez speculated, left only Francesca. . . .

Without any great confidence he visited the hut in search of clues. After several hours of fruitless rummaging, he noticed a brownish mark on the bedroom door, suddenly made visible by the evening sun. Closer inspection revealed it to be the imprint of a bloodstained human thumb. Alvarez obtained a saw, cut out that section of the door, and returned to the police station with orders that Francesca be brought in. He rolled her right thumb on an ink pad, then pressed it upon a sheet of paper. Even to Alvarez's relatively inexperienced eye, the imprint when studied under a magnifying glass matched the bloody thumbprint on the door.

Confronted by the evidence of her own hand, Francesca's previously steely resolve collapsed. She admitted killing her children because they blighted her chances of marriage to the young man. After battering them to death with a stone, she had tossed the stone into a well and washed her hands carefully. But inadvertently she had touched the bedroom door.

Francesca Rojas was convicted and sentenced to life imprisonment, her place in criminal history assured.

Conclusion

For Juan Vucetich, the events at Necochea were a glittering triumph. He wrote to a friend, "I hardly dare to believe it, but my theory has proved its worth. . . . I hold one trump card now, and I hope I shall soon have more." Sadly, his confidence was misplaced, as official hostility to fingerprinting forced a return to Bertillonage. Undismayed, Vucetich fought back and in 1916 actually succeeded in setting up a General Register of Identification predicated on fingerprinting the entire population. But the scheme proved hopelessly unpopular. Riots and civil disorder eventually forced the government to abandon the plan and destroy its records. In 1925, Juan Vucetich died—some said of disillusionment.

The Stratton Brothers

DATE: 1905

LOCATION: London, England

SIGNIFICANCE: Although the first British trial to admit fingerprint evidence occurred in 1902—a burglar was convicted of stealing billiard balls—it was not until this sensational murder that the new technique made headlines.

For more than two decades, Thomas Farrow and his wife, Ann, had managed a paint shop in Deptford, South London. The elderly couple were popular in the neighborhood and were regular in their habits, which explains why on Monday, March 27, 1905, puzzled because the store had not opened for business as usual, other employees

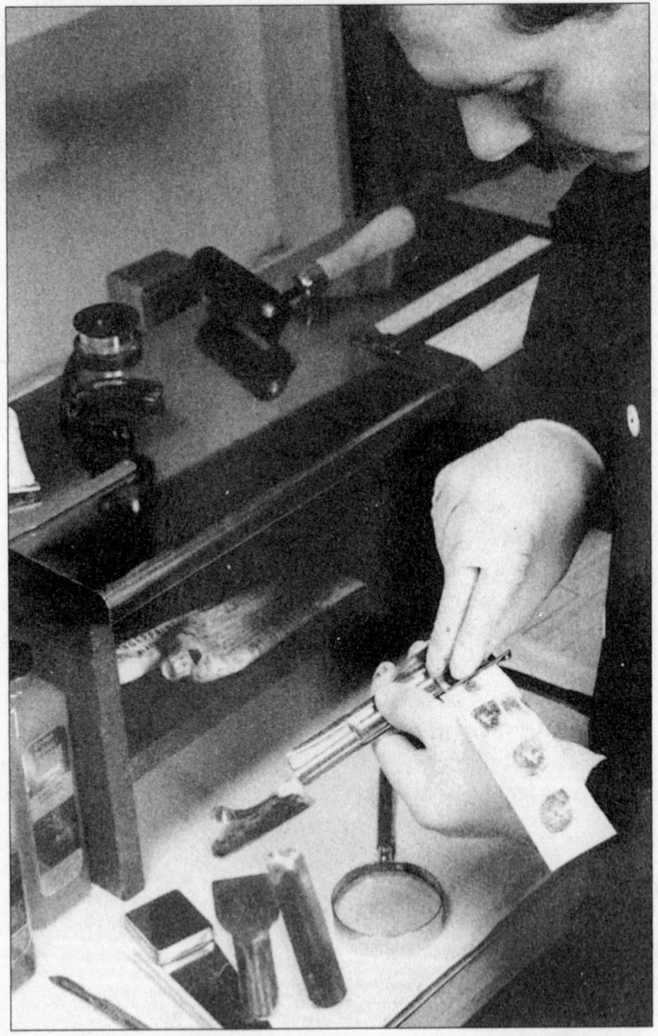

Fingerprint expert obtaining fingerprint specimens. (FBI)

forced an entry. The reason for the lapse was soon apparent. In the parlor, Mr. Farrow had been bludgeoned to death; upstairs his wife lay sprawled across a bloodstained bed, still alive but just barely. She, too, had been savagely beaten. Four days later, without ever regaining consciousness, Ann Farrow died.

It was common knowledge locally that every Monday the Farrows handed over the week's earnings to the store owner, George Chapman, so there was little doubt that robbery had been the motive. An empty cash box and its inner tray lay on the bedroom floor, and there was a mask fashioned from a black silk stocking. The discovery of two similar masks downstairs suggested the presence of more than one intruder. The cash box and tray were examined by Scotland Yard's Fingerprint Branch. They found a clear print on the tray that did not correspond to either of the victims or any of the police officers present (one of whom admitted to having touched the tray). Neither did it tally with any of the eighty thousand prints already on file at the Yard.

Milkman Henry Jennings and his young assistant told of seeing two men leave the store at about 7:15 on the morning of the murder. They described the taller of the two as dressed in a blue serge suit and bowler hat, and walking hurriedly. His companion was wearing a dark brown suit, cap, and brown boots. Neighborhood rumor soon put names to these descriptions—Albert and Alfred Stratton, two brothers with a record of burglary who had been seen in the vicinity of the crime scene. Albert's landlady, when interviewed, recalled him asking if she had an old pair of stockings she could give him, to which she replied no. Later, while making her lodger's bed, she found a silk stocking mask hidden between the mattresses.

Yet more circumstantial evidence was supplied by Hannah Cromerty, Alfred's mistress. In no mood to help her violent lover—she was sporting a black eye—she revealed that he had been out on the night of the murder; the next day he had given away his overcoat and blacked over the brown shoes he'd worn.

Brothers Caught

On Sunday, after a week spent scouring the brothers' known haunts, detectives tracked Alfred to a public house. His brother, he claimed, had gone to sea, but the next day Albert was located in a nearby lodging house. Both were taken to Tower Bridge police station, where Detective Inspector Charles Collins, a founding member of Scotland Yard's Fingerprint Branch in 1901, rolled their fingers on the ink pad. He took impressions of every digit, then compared them to the print on the tray. It matched Alfred's right thumbprint exactly. As a consequence, both brothers were charged with murder.

At their trial, the prosecution, led by Richard Muir, suffered a severe setback when neither milkman could identify the accused as the men they had seen leaving the store. This shifted the burden of the

Crown's case entirely to the fingerprint evidence. With the aid of photographic enlargements, Collins demonstrated twelve points of resemblance between Alfred's thumbprint and the impression left on the tray. Just as significant was the complete absence of characteristics that did not agree. The jury was clearly impressed, watching avidly as Collins, asked to clarify the technique of obtaining prints, fingerprinted one of their own members.

Quite naturally, the defense dismissed any suggestion that fingerprinting was an infallible means of identification. To argue this point, they produced Dr. John Garson, longtime advocate of Bertillonage and occasional practitioner of the new technique. After a blistering personal attack on Collins, the doctor went on to state categorically that the thumbprint on the cash box did not belong to Alfred Stratton. At this, Muir produced two letters, both written by Garson on the same day, offering his services to both the prosecution and the defense. Such opportunism prompted a caustic response from the judge, who made his own feelings about Garson's duplicity abundantly clear. Even so, the judge stressed that the thumbprint evidence should be approached circumspectly, but he did admit the extraordinary resemblance. What weight the jury gave this admonition is unknown; what is certain is that they found both Strattons guilty of murder.

Conclusion

Besides ruining his own career, Garson's recklessness did have one other interesting outcome. Thereafter, the medical profession rarely concerned itself with fingerprinting; it became, and remains, the domain of the skilled specialist. As for the Strattons, their unwitting contribution to the advancement of forensic science far outlived them. Both were hanged on May 23, 1905.

Thomas Jennings

DATE: 1910

LOCATION: Chicago, Illinois

SIGNIFICANCE: History was made when this case resulted in the first American murder trial to admit fingerprint testimony.

Clarence Hiller lived with his wife and four children in a two-story house on West 104th Street in Chicago. In the early hours of September 19, 1910, Mrs. Hiller awoke and noticed that a gaslight near the bedroom of their daughter Florence was not burning as it should. She aroused her husband and he hurried to investigate.

On the landing he encountered a stranger. In the ensuing struggle, both men tumbled down the staircase. Moments later, Mrs. Hiller heard two shots and then her husband's faint call for help. This was followed by the slamming of the front door. At the foot of the stairs, Mrs. Hiller found the lifeless body of her husband.

A neighbor, John C. Pickens, alerted by the shooting and the woman's screams, came on the double. By chance, his son Oliver had been talking to a policeman, Floyd Beardsley, just a short distance away. Both heard the shots and arrived immediately. Beardsley heard how the murderer, bearing a lighted match, had entered the bedrooms of Hiller's daughter Clarice, fourteen, and her sister Florence, a year younger. Neither girl was assaulted, but both were too terrified to cry for help. Just moments later the fatal struggle began.

At the foot of Florence's bed were some particles of sand and gravel. Together with three unused cartridges found near the body, they seemed to be the only clues. However, unknown to Beardsley, the killer was already in custody, having been arrested on wholly unrelated charges.

Less than a mile from the Hiller residence, four off-duty officers had spotted a man darting through the darkness, glancing furtively behind him, as if fearful of pursuit. Told to empty his pockets, he handed over a loaded .38 revolver. He gave his name as Thomas Jennings. There were fresh bloodstains on his clothing, and he had injured his left arm, both the result, he said, of having fallen from a streetcar earlier that night.

Unimpressed by this story, the officers escorted Jennings to the station, where news of the Hiller murder had just come in. Immediately the wounded man became a prime suspect, especially when it was learned that only a month earlier he had been freed from Joliet State Penitentiary after serving time for burglary.

Suspicion began to accumulate. Just two weeks after being released, under an assumed name and in violation of his parole, Jennings had purchased the .38 revolver. When examined, its cartridges were found to be identical with those lying next to Clarence Hiller's body.

Prints in the Paint

The most significant discovery occurred at the murder scene. Next to the rear kitchen window, through which the killer had effected his entrance, were some railings. By chance, Mr. Hiller had painted these very railings just hours before his death, and etched into the fresh paint were four fingerprints from someone's left hand.

Ever since the 1905 visit of Scotland Yard's Sergeant John Ferrier to the St. Louis International Police Exhibition, at which he extolled the virtues of Sir Edward Henry's classification system, American law enforcement agencies had adopted the principle of fingerprinting with varying degrees of enthusiasm. As late as 1910, Bertillonage still held sway with most police forces, but Chicago had embraced the new technology with open arms. Technicians from the Police Department Bureau of Identification (PDBI) studied the railings and made enlarged photographs of the prints. When compared to those of Jennings, they matched in every detail.

At the ensuing trial, a quartet of fingerprint experts headed by William M. Evans of the PDBI agreed that Jennings's hand, and his alone, had left the prints on the railings. Their testimony led to a verdict of guilty and a sentence of death. Immediately defense lawyers filed an appeal on the grounds that fingerprinting was still a fledgling discipline, unworthy of the approbation "scientific," and that Illinois laws did not recognize such evidence.

Across the nation, eyes were on the Illinois Supreme Court, awaiting its decision—which came on December 21, 1911. In a historic ruling the court first sanctioned the admissibility of fingerprint evidence and then upheld the right of experienced technicians to testify as experts, saying:

> When photography was first introduced it was seriously questioned whether pictures thus created could properly be introduced in evidence, but this method of proof, as well as by means of X-rays and the microscope, is now admitted without question. . . . We are disposed to hold from the evidence of the four witnesses who testified, and from the writings we have referred to on this subject, that there is a scientific basis for the system of fingerprint identification, and that the courts cannot refuse to take judicial cognizance of it. . . . If inferences as to the identity of persons based on voice, the appearance or age are admissible, why does not this record justify the admission of this fingerprint testimony under common law rules of evidence?

In conclusion, the appeal court affirmed the verdict of the jury and upheld Jennings's death sentence. He was later hanged.

Conclusion

Boosted by the Illinois decision, fingerprinting soon reached every corner of America. In 1915, a small group of fingerprint experts met in California and founded what was later called the International Association for Identification (IAI), the world's first such professional body. A gathering house of information and new developments, the IAI today has thousands of members worldwide.

William Berger

DATE: 1920

LOCATION: Berkeley, California

SIGNIFICANCE: As criminals in the United States became more mobile, agitation for a national fingerprint register acquired momentum.

One of the most peculiar crime waves ever to hit California began in Berkeley in September 1920, when a thief sawed his way into a grocery store, helped himself to some cookies, drained a bottle of milk, and left. An otherwise nondescript offense was made somewhat less so when, among fingerprints found at the scene, detectives noted one in the shape of a question mark. At that time, the Berkeley police department was headed by August Vollmer, an innovative crime fighter who would later figure prominently in the development of the polygraph. Under his auspices, the recently founded fingerprint department had thrived, but on this occasion it was unable to provide a match for the unusual print.

Two weeks later, the burglar struck again, this time at a hardware store. He used the same method of entry, left the same distinctive fingerprint, and again demonstrated his fondness for milk. All that winter, Berkeley suffered a plague of minor robberies. Nothing much was ever taken, but not a single item was ever recovered. Police, used to the proceeds of petty theft ending up in pawnshops and the like, were puzzled. And then, in April 1921, the break-ins abruptly ceased. A month, then two, passed without incident. Gradually memories of the so-called Question Mark Burglar began to fade.

The following September, though, soon brought them back into sharp relief, as a new outbreak of nocturnal burglaries left little doubt that the idiosyncratic thief was back. There was no mistaking the pattern: always between 9 P.M. and 4 A.M., always in the same small geographic area, always the same modus operandi, and almost always interrupted with a refreshing bottle of milk. As before, the robberies ceased in April, only to recommence in September, then halt again in April 1922. With the regularity of migrating birds, the following September heralded the arrival of another flock of burglaries.

As the decade wore on, the six-month cycle remained constant: arrive in September, leave in April. Once, the elusive thief was nearly arrested. A police officer, spotting someone sneaking away from a grocery store under cover of darkness, shouted for him to stop. Instead, he took to his heels. The ensuing chase, punctuated with bullets from both parties, resulted in the burglar being hit but still managing to escape. Local hospitals and doctors were alerted to look out for a man with a bullet wound, but no report ever came in.

Rich Pickings

By 1928, the previously petty criminal was stealing goods worth thousands of dollars each winter, and apart from the fingerprint, detectives had only one lead. Because of how and where he broke into each house—always at the point of easiest access in the wall, and always neatly—it was assumed that he was either a carpenter or a plumber. Vollmer ruminated. Where did skilled artisans go in the summer?

Eventually inspiration struck. Each spring, a fleet of fishing boats left San Francisco's Bay Area and headed north to the Alaskan salmon fishing grounds. Besides the hands, all carried expert workers who could make running repairs. The Berkeley agency that sent crews to Alaska compiled a list of thirty-eight names and addresses, which was forwarded to the FBI.

In 1924, the FBI had received 810,000 sets of fingerprints from the Bureau of Criminal Investigation at Leavenworth. These formed the core of what would become a national system of fingerprint identification, planned and instituted by J. Edgar Hoover. America's burgeoning criminal population was taking to the highways, the railroads, even the air, in their attempts to fatten bankrolls and outwit pursuers. Crimes separated by hundreds or even thousands of miles were rarely connected, and hoodlums could, with impunity, leave their fingerprints at crime scenes, secure in the knowledge that chances of capture were negligible. Hoover was out to change all that. He envisaged a national fingerprint system that would make a criminal's record available to every police force in the country. Vollmer had an early example of what lay ahead; what he hadn't been able to uncover in eight years, the FBI found in days—the name of the Question Mark Burglar.

William Berger was indeed a ship's carpenter; he was also a thief who had been fingerprinted while serving time at San Quentin in 1914. The highly individual print came from the middle finger of his left hand. Tempering Vollmer's elation at discovering his identity was the realization that this was July, and records showed Berger as being somewhere in Alaska, well beyond anyone's jurisdiction.

Suspect Returns

All they could do was wait. Sure enough, that September, William Berger returned to his house on Stannage Avenue, right in the heart of the burglary catchment area, unaware that his every move was now the subject of intense scrutiny. One night, as the officers watched, Berger crept out of the house in the early hours. Rather than attempt to tail him in the fog, they decided to await his return. At 4 A.M. their vigil ended. Berger loomed through the murk, laden down with booty. Just as he was about to enter his home, the officers challenged him. Instantly he fled. Several blocks later, having ignored repeated warning shots, Berger was struck by a bullet and fell dead to the ground. When

J. Edgar Hoover, soon after his appointment as Director of the FBI. (National Archives)

inked at the morgue, the middle finger of his left hand revealed the telltale print that had eluded the Berkeley Police Department for so long.

But William Berger still had one more secret to divulge. Every item he had ever stolen was found stashed within his house—the proceeds, it was estimated, of more than four hundred robberies. Financial gain had never been a consideration; a true kleptomaniac, he stole because he couldn't stop. His wife, terrified of his maniacal rages, had been too afraid to inform the authorities. She confirmed that Berger had indeed been struck by a bullet all those years before, but again she had not been able to bring herself to call the police. Humanely, it was decided that the woman had suffered enough, and all charges against her were dismissed.

Conclusion

Gradually Hoover's dream of a national fingerprint reference library became a reality. By 1990, the FBI had more than eighty-three million fingerprint cards on file, representing more than twenty-two million people. Each day, another twenty-seven thousand cards are added to the system.

The Kelly Gang

DATE: 1933

LOCATION: Oklahoma City, Oklahoma

SIGNIFICANCE: In this extraordinary case, the victim's own fingerprints, not those of his captors, led to multiple convictions for kidnapping.

On the evening of July 22, 1933, a bridge game between Oklahoma City oil millionaire Charles Urschel, his wife, and their friends the Jarretts on the patio of Urschel's home was abruptly interrupted by the sudden appearance of two gunmen. "Which of you is Urschel?" snarled one of the bandits. When neither man at the card table responded, both were forced into a waiting car and driven off. Although ordered not to do so, Mrs. Urschel quickly called the police and the FBI. Within ninety minutes, Walter Jarrett was back at the Urschel house, shaken but otherwise unharmed. Once the kidnappers had clarified his identity, they had taken his wallet, pushed him from the car, and then driven off with Urschel.

Four days later, a ransom note arrived in the mail. Besides a demand for two hundred thousand dollars, the kidnappers enclosed a note in Charles Urschel's handwriting. To signify their acceptance of the terms, the family was told to place an advertisement in the classified column of the *Daily Oklahoman* reading: "FOR SALE—160 acres land, good five room house, deep well. Also cows, tools, tractors, corn and hay. $3,750 for quick sale. TERMS. Box H-807."

Soon afterward, another family friend, E. E. Kirkpatrick, received fresh instructions to check into a Kansas City hotel. Later, carrying a bag stuffed with two hundred thousand dollars in marked bills, he left the hotel as arranged and was approached by a tall stranger "in a natty summer suit with a turned down Panama hat." The man took the grip, assured Kirkpatrick that Urschel would be returned safely, then disappeared. Twenty-four hours later, exhausted, Urschel arrived home.

Extraordinary Detail

Charles Urschel was clearly a remarkable man. The account he gave FBI agents of his ordeal is unparalleled in the annals of kidnapping: Every detail, every incident, no matter how mundane, had been consciously logged for possible future use as evidence. After releasing Jarrett, the kidnappers had blindfolded Urschel and driven through the night. At dawn, they switched cars in a garage or barn. Urschel was told to lie down on the rear floor of a large car he guessed to be either a Buick or a Cadillac. Three hours later, they pulled into a gas station. As a woman filled the tank, one of the kidnappers idly asked about local farming conditions. The woman replied that crops in that area were "all burned up."

At the next stop, another garage or barn, Urschel overhead one gang member mention the time—2:30 P.M. After dark, he was taken on foot to a house where he spent the night. The next morning, the final leg of his journey ended at a farmhouse surrounded by cows and chickens.

Despite being blindfolded and handcuffed to a chair, Urschel missed nothing. When water was being drawn from the farmhouse well, he noticed that the windlass creaked. And the water, drunk from a tin cup without a handle, had had a distinctive mineral taste. By loosening his blindfold a fraction, he was able to glimpse his watch. He noted that each morning at 9:45 and each evening at 5:45 a plane passed over the house. But on Sunday, July 30, there was a downpour of rain, and he didn't hear the morning plane.

Neither did Urschel limit his detective talents to mere observation. While writing the letter to his family, unnoticed by his kidnappers, he deliberately left fingerprints on every surface he could reach. On July 31, his ordeal came to an end when he was driven to Norman, on the outskirts of Oklahoma City, and released.

Wide Search

FBI efforts to locate the kidnappers' lair centered on the rainstorm and the plane's failure on that Sunday to follow its usual flight pattern. All airlines that flew within a six-hundred-mile radius of Oklahoma City were contacted for details of their schedules. Meanwhile, the comment by the gas station attendant prompted calls to meteorological offices to ask if any drought-ridden regions had recently received a heavy drenching.

American Airlines reported that on Sunday, July 30, a plane on the Fort Worth–Amarillo run had been forced to swing north from its usual course to avoid a heavy storm over an area previously afflicted by scorching drought. Calculations showed that the morning plane and the return afternoon flight would pass over Paradise, Texas, at the approximate times recalled by Urschel.

Posing as bankers seeking to extend loans to farmers, FBI agents began visiting every farm in the area. Eventually they came to the five-hundred-acre ranch of Mr. and Mrs. R. G. Shannon. Here they arrested Harvey Bailey, a notorious hoodlum, with seven hundred dollars of the marked ransom bills on him. Then investigators remembered that the Shannons' daughter Kathryn was married to George "Machine Gun" Kelly, an infamous if somewhat reluctant gangster (his notoriety was almost entirely due to his wife's mythomania). Warrants were issued for the arrest of anyone connected with the Kellys.

When shown the Shannon home, Urschel immediately identified it as his place of capture. There was the well, the old tin cup, and the chair to which he had been handcuffed. And he could never forget the taste of that water. Most conclusive of all, however, were Urschel's fingerprints; according to one expert, they covered almost every square inch of reachable surface.

Eventually the entire Kelly gang was rounded up and given long jail terms.

Conclusion

It seemed almost inevitable that "Machine Gun" Kelly, a hapless bungler who never once fired a gun in anger, should have chosen

Charles Urschel, possibly the most astute hostage ever, for his one excursion into kidnapping. In such a lopsided contest, Kelly never stood a chance. He spent the rest of his life behind bars and died in Leavenworth in 1954.

Edward Morey

DATE: 1933
LOCATION: Wagga Wagga, Australia
SIGNIFICANCE: The durability of fingerprints is well documented—Egyptian mummies have been found with their prints intact—but here a corpse yielded its secrets in a truly bizarre fashion.

One of Australia's most famous murder cases began on Christmas Day 1933, when fishermen on the Murrumbidgee River, close to Wagga Wagga, found a body snagged on a tree. At the autopsy, it was determined that the corpse was male, strongly built, aged between forty-five and fifty; had received several crushing blows to the back of the head; and had been in the water for about five weeks. Identification would not be easy. The features were unrecognizable, and there was no possibility of fingerprints: The right hand was missing completely, and the left was badly mangled.

Under the direction of Police Commissioner Walter H. Childs, detectives began searching the riverbank for clues. Before long, one of them noticed something brownish and shriveled-looking caught in the bushes. Wading into the water, he retrieved the object and held it up for closer examination. It took a moment to realize what it was.

Back at the laboratory, experts confirmed the find—human skin with a thumbnail still attached. Actually, it was the outer skin of a right hand and wrist, all of the flesh having been eaten away by maggots. What remained looked like a dirty glove—which gave the technicians an idea.

If properly treated, the skin might yet yield a usable fingerprint, but somehow they had to find a way to take that print. The answer was gruesomely simple. First, they needed someone whose hand roughly conformed to the size of the skin. The officer who most closely met this requirement then put on a surgical glove and slid his own right hand into the dead man's skin. It was a delicate operation; one careless move might tear the skin apart and destroy any chance of success. Once the detective's fingers had filled out the limp skin, he pressed the digits on an inked pad and then onto a fingerprint card. The result was a surprisingly clear set of prints with identifiable loops, whorls, and arches. Encouraged by this success, the experts managed to peel the skin off the mangled left hand and obtain a set of prints from it. Now began the task of searching thousands of fingerprint records, with no guarantee that their target was among them.

Hobo Killer

Their good fortune held. The dead man was Percy Smith, an itinerant who traveled the Wagga Wagga region in a dilapidated wagon, taking odd jobs wherever he could. Locals confirmed that Smith had been missing for several weeks. They also mentioned that he and a fellow drifter, Edward Morey, had been seen together near where the body was found. Later Morey was known to have sold some of the dead man's tools.

A widespread manhunt was abruptly halted when officers were told that Morey was already in jail on vagrancy charges. His angry denials of complicity in Smith's death rang hollow when confronted by the evidence. Besides the tools, he had also sold Smith's monogrammed gold watch, and when investigators searched his camp, they found a bloodstained ax that he attempted to conceal from them.

News of the unique fingerprint retrieval method caused a sensation at Morey's trial and dominated every headline until the chief prosecution witness, Moncrieff Anderson, was suddenly gunned down at his home. His wife, Lillian, blamed the killing on intruders until the police noticed that Mrs. Anderson's handwriting bore an uncanny resemblance to that of "Thelma Smith," authoress of two love letters found in Morey's possession. The shooting in no way affected Morey's fate; he was convicted and sentenced to hang, though later reprieved. After serving twenty years, he was freed on health grounds and shortly thereafter died from tuberculosis.

Conclusion

At her own trial, Lillian Anderson, a confused and tragic figure with a mental age of fourteen, claimed that her dead husband and not Morey was the killer. Prosecutors soon established that she had shot her husband and named him as Smith's murderer in order to save her lover. Morey showed no such loyalty. In his testimony, he turned his back on the woman who had tried to save him, denying that he even knew her. Lillian Anderson's conviction for manslaughter earned her a twenty-year jail sentence.

Peter Griffiths

DATE: 1948

LOCATION: Blackburn, England

SIGNIFICANCE: For the first time, mass fingerprinting of the population was used to bring a savage killer to justice.

At 1:20 A.M. on May 15, 1948, a nurse making the rounds of the children's ward of the Queens Park Hospital, near Blackburn, Lancashire, realized that three-year-old June Devaney was missing from her cot. An hour earlier, the toddler had been sleeping peacefully, recovering from pneumonia, and was due to go home later that day. After fruitlessly searching the immediate vicinity, the hospital staff contacted the police. At 3:17 A.M., a constable found June's body lying beside the hospital wall. Her horrific injuries were described by the police surgeon as follows:

> There was a multiple fracture of the skull. . . . Blood was exuding from the nose. . . . There were several puncture wounds on the left foot. They might easily have been caused by fingernails gripping the left ankle. . . . The injuries to the head were consistent with the head having been battered against a wall. It appears to me that the child was held by her feet and the head swung against the wall.

In addition, the child had been raped, and there were teeth marks on her left buttock.

Footprints on the ward's polished floor intimated that the killer had prowled between the cots in stocking feet. He also appeared to have moved objects around. A bottle of sterile water—normally kept on a trolley at the end of the ward—now stood under June's cot. The nurse was adamant that on her earlier round the bottle had been in its rightful place.

For fifteen hours, Detective Inspector Colin Campbell, chief of Lancashire's Fingerprint Bureau, scrutinized the ward. He began with the footprints: ten and a half inches long, almost certainly male, unlikely to have been made by a member of the nursing staff. Next came the misplaced bottle. Judging from the myriad fingerprints, it had been handled by several different people. By that evening, every member of the hospital staff had been fingerprinted for comparison with those on the bottle. All except one set of prints were accounted for; Campbell had little doubt that these belonged to the killer.

When records also failed to make a match with any known criminal, investigators reconciled themselves to the likelihood of a protracted inquiry. They began by compiling a list of all people with general access to the hospital—2,017 people in all, of whom 642 had specific access to the children's ward. All were fingerprinted, yet none matched the mystery prints left on the bottle.

With his options fast diminishing, the commanding officer, Detective Inspector John Capstick, embarked on an unprecedented step. Because the hospital grounds were difficult to negotiate after dark, he believed that the killer was most probably a man with local knowledge. Therefore, he proposed that every male over age sixteen in the town of Blackburn be fingerprinted. It was a revolutionary and daunting prospect. The electoral register showed no less than thirty-five thousand houses; all would have to be visited.

Historic Sweep

The immense operation began on May 23, with officers going from house to house, fingerprinting every male who had been in Blackburn on the night of the murder, recording each set of impressions on a small card. Despite the response—more than forty thousand males—by mid-July the investigators were no nearer to identifying the print. With the electoral register exhausted, it was beginning to look as if the killer had slipped through the net. And then inspiration struck. Because of food shortages since the war, every British adult carried his or her own rationing book; possibly names might crop up there and nowhere else. Three weeks of cross-checking ration book records against the electoral register showed at least two hundred males still unaccounted for.

One of these was Peter Griffiths, a twenty-two-year-old flour mill packer who lived at 31 Birley Street, Blackburn. On August 11, officers called and asked if they could take his fingerprints. Griffiths, whose niece had been in Queens Park Hospital at the same time as June Devaney, agreed. Later that day, his card—number 46,253—was routinely filed with the bureau. The next afternoon, one of the searchers suddenly cried out, "I've got him! It's here!"

At first Griffiths denied the crime; then he made a confession, saying, "I hope I get what I deserve." On the night in question, having consumed a large amount of alcohol, he had gone to the hospital, a

Fingerprint expert comparing latent fingerprints developed during an investigation. (FBI)

place he knew well from his childhood. Leaving his shoes outside, he crept into the children's ward. He claimed that a child woke up when he stumbled against her cot and that he carried her outside to keep her silent. But when she wouldn't stop crying, "I lost my temper . . . and you know what happened then." An already strong forensic case against Griffiths was further buttressed by the discovery of fibers from the victim's nightdress on his suit, while his feet fitted exactly those prints found on the hospital floor.

Rejecting a defense plea of diminished responsibility, the jury found Griffiths guilty of murder, and on November 19, 1948, he was executed.

Conclusion

Right from the outset, the police moved to allay concerns about privacy by announcing that all prints unconnected with their inquiry would be publicly destroyed. This was a promise kept, as was their assurance that no one would be compelled to have his prints taken. As it turned out, public revulsion at the ghastliness of June Devaney's death ensured that objections to the exercise were few and refusals virtually nil.

George Ross

DATE: 1951

LOCATION: Cleveland, Ohio

SIGNIFICANCE: Although most fingerprints are of the latent type—invisible until dusted—here the killer unwittingly recorded his prints for posterity.

While cruising Cleveland's east side on December 8, 1951, patrolman Forney Haas spotted a Lincoln with California plates driving in the wrong direction down a one-way street. After a brief chase, the Lincoln drew to a halt. The driver apologized for his unfamiliarity with the local traffic system but could not produce his license when asked. It was, he said, back at his boardinghouse. Haas insisted on seeing it, and the two cars set off in convoy. When they reached the rooming house at 8210 Euclid Avenue, both men went indoors.

A short while later, as she was cleaning the hallway, the landlady, Lottie Cooper, heard male voices arguing. One shouted, "I'm telling you my license was in my wallet when I left. It's been stolen from here!" A fusillade of bullets terminated the outburst. The door flew open, and a man brandishing a revolver ran past Mrs. Cooper and out into the street. Haas, grievously wounded, was rushed to the hospital, but he died half an hour later without regaining consciousness.

A tearful Mrs. Cooper told officers that the gun-wielding fugitive had given his name as Montgomery when renting the room. But his true identity was revealed when detectives ran a check on the abandoned Lincoln. It had been stolen in California and used by two criminals in a string of robberies across the country. One of the bandits was thought to be George Ross, a twenty-seven-year-old hood who had graduated from car theft to grand larceny. When shown a picture of Ross, Mrs. Cooper identified him as the slim, swarthy man she knew as Montgomery. Immediately a description was circulated to police nationwide.

Meanwhile, Ross had fled in a panic after the shooting. Eventually he found his way to Bedford, a suburb of Cleveland, where he stopped at a diner to take stock of his situation. He needed to get rid of Haas's revolver, the murder weapon; its serial number would be on record, and worse, his fingerprints were all over it. Earlier in his career, Ross had neglected to wipe his prints off a pistol. That lapse had resulted in a lengthy prison term; any similar blunder this time would mean the electric chair. As he pondered this unappetizing prospect, an idea came to him, one that would get rid of the gun for good.

Ditched Weapon

He went to the restaurant washroom and locked himself in a cubicle. Once the washroom was empty, he dropped the revolver into the toilet bowl and flushed it from view. Afterward Ross stole a parked car and headed south. But he was already a marked man; a bystander had seen him take the car and called the police.

For the next week, Ross dodged from one state to another, always staying one step ahead of his pursuers. He attempted to rob a gas station in West Virginia but had to flee when a customer arrived unexpectedly. In his haste to escape, Ross dropped a revolver, but early official elation at this discovery evaporated when tests eliminated it as the murder weapon. Two days later, he was asleep in a car in Ellicott City, Maryland, when two police officers approached. In the brief gun battle that ensued, Ross was hit but managed to escape.

His luck finally ran out on December 17. By now, radio bulletins, including his description, were a regular feature on the daily news programs, so it was no surprise when a farmer in Maryland, Edward Duval, recognized the unkempt hitchhiker he had picked up as Ross. Wisely, Duval said nothing until he had dropped Ross off. Then he found two police officers. When challenged, Ross insisted that his name was Perkins, but exhausted from days of running, he soon admitted his identity and went meekly into custody.

When he came to trial on February 4, 1952, Ross claimed that he had killed Haas in self-defense. It was a weak argument, and, anticipating its rejection by the jury, the defense team cobbled together an alternative and uncomplimentary insanity plea for their client, describing him as a "moral imbecile" and "psychopathic." But one piece of evidence in the prosecution's arsenal proved devastating.

Shortly after the Cleveland shooting, a customer at the diner in Bedford reported that one of the toilets was backed up. A friend of the proprietor's volunteered to clear the obstruction. After a few moments' tentative exploration, he found the source of the trouble. Lodged in the waste pipe, just around the bend, was the missing revolver. Its serial number, when checked against the police records, confirmed that this was the gun issued to Officer Forney Haas.

After several days of immersion in water, there was little reason to expect that the gun would yield any identifiable fingerprints. Sure enough, the handle and trigger were clean, but when technicians looked more closely at the barrel, they found to their astonishment a clear fingerprint. More amazing still, the print had seemingly been etched into the gunmetal. Experts studied the print in astonishment: Its like had never been seen before.

Various theories were advanced to explain this extraordinary phenomenon. Some attributed it to the heavy lime content of the water, or a chemical reaction with the copper ball in the tank; others thought that Ross's own sweat, which contained an unusually high salt content, had eaten into the metal of the gun as he handled it.

Whatever the reason, the revolver proved decisive. Introduced right at the end of the prosecution's case, it caused a sensation and left the defense reeling. Convicted of murder, with no recommendation for mercy, Ross went to the electric chair on January 16, 1953.

Conclusion

Fingerprints have immense durability. Under optimum conditions they can remain intact on a surface for decades (see the Valerian Trifa case on page 136). Although the circumstances of George Ross and the corrosive fingerprint can only be described as freakish, they remind us that every time we touch something, until that surface is either cleaned or degraded, our presence at that spot has been indelibly recorded.

Richard Ramirez

DATE: 1984

LOCATION: Los Angeles, California

SIGNIFICANCE: Computerization of the Los Angeles Police Department (LAPD) fingerprint records dealt a deadly blow to one of California's worst killers.

The man who became known as the Night Stalker first struck in June 1984, when a young woman was stabbed to death in her Los Angeles home. Over the course of the next year, another twelve people would die at the hands of this psychopath. The murders followed a set pattern, usually occurring in the early hours of the morning and most often at homes where a window was left open. After gaining entry, the killer would sever the phone lines and then go about his evil business. Any adult males present were dispatched with a bullet to the head. Female victims were raped, often next to the bodies of their dead partners. At the seventh murder scene, the killer daubed a pentagram, a sign often associated with devil worship, on a bedroom wall and on the thigh of one female victim. Another woman, after seeing her husband shot to death, was ordered to "swear upon Satan" while being sodomized. Survivors described their assailant as a tall, slim, scruffy Hispanic male, unwashed and stinking, with bad teeth.

The homicidal outburst culminated at Mission Viejo in August 1985, when a young couple was attacked at home. The Night Stalker shot the man through the head while he slept, then raped the woman beside him. Somehow, in the midst of this nightmare, the woman managed to keep her wits about her enough to note the license plate number of the killer's old orange Toyota as he drove off. Records showed that the car had been stolen in Chinatown while the owner was eating in a restaurant. An all-points bulletin led to the vehicle being located two days later in a parking lot. After mounting around-the-clock surveillance in case the killer returned—he didn't—officers eventually removed the car for forensic examination. This proved vital, as experts managed to lift a partial fingerprint.

In a city the size of Los Angeles, manually searching fingerprint files was a tedious process that could take days, and even then, human error always left open the possibility of missed correlations. But in

1985 the LAPD had installed a computerized fingerprint database system similar to that used by the FBI and capable of more than sixty thousand comparisons per second. The system works by storing information about the relative distance between different features of a print

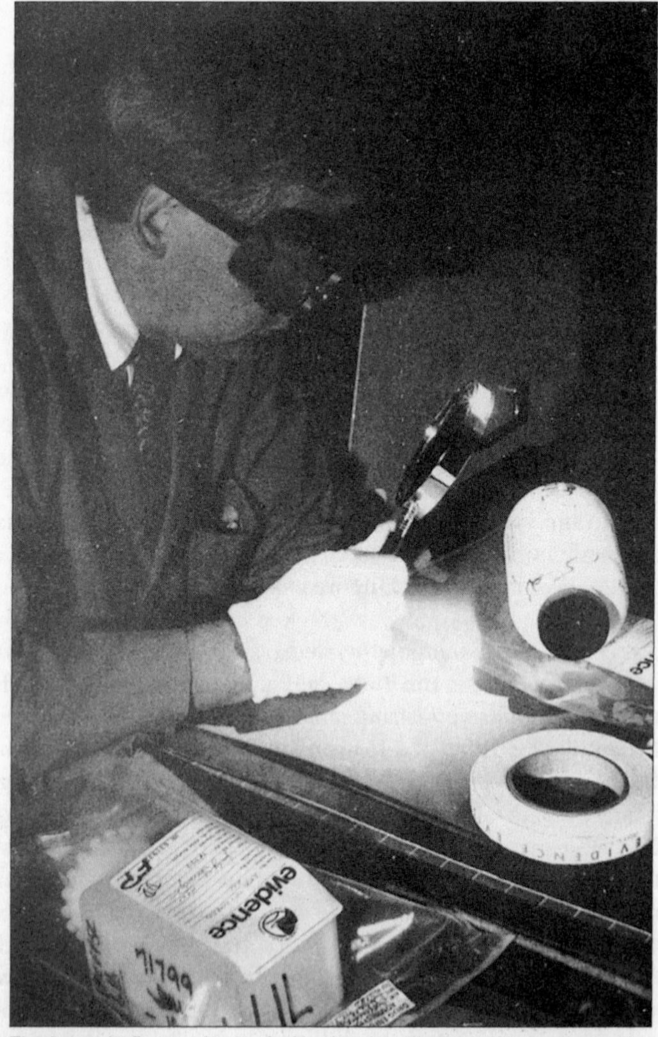

The latest in fingerprint analysis—laser technology that seeks out latent prints. (FBI)

and comparing them to a digitized image of the suspect's fingerprint. Within minutes, the computer provided a positive match for the print from the orange Toyota—the Night Stalker was Richard Ramirez, a twenty-five-year-old drifter from El Paso, arrested several years previously on a misdemeanor traffic violation.

Forensic Miracle

Those working in the fingerprint department described the identification as "a near miracle." The computer had only just been installed, and this was one of its first trials. Moreover, the system contained the fingerprints of only those criminals born after January 1, 1960—Ramirez was born in February 1960! Immediately his name and photograph were circulated to the media. Ramirez, on his way back from Phoenix, Arizona, after buying cocaine, returned to Los Angeles unaware that his face was front page news.

Customers in the East L.A. liquor store first recoiled from the foul-smelling man buying a Pepsi and doughnuts and then noticed his similarity to the wanted killer whose picture appeared on the newspaper rack. East L.A. is a predominantly Hispanic area, and locals, infuriated by the prejudice that the Night Stalker had aroused against them, were in no mood to apprehend Ramirez gently. A wild chase ensued. After trying unsuccessfully to steal several cars, Ramirez ran panting into the arms of a waiting patrolman (curiously, also named Ramirez) and begged to be taken into custody before the crowd lynched him. A few days later, he was arraigned on thirteen counts of murder.

Ramirez, a master manipulator, managed to delay his trial for more than three years. During this time, he changed his appearance drastically, washing regularly, wearing smart clothes, and having extensive dental work done so that the man in court bore little resemblance to published descriptions of the Night Stalker. Even so, the case against him was ironclad. The gun used in the killings—a .22 caliber semiautomatic—was traced to a friend's house, and jewelry stolen from the victims turned up at the home of his sister.

Another year would pass before he was convicted. In that time, hardened police and court officials stated that no defendant had ever so unnerved them. Ramirez, menacing behind dark glasses, would erupt into totally unpredictable paroxysms of rage, invoking Satan's name and cursing everyone around him. On November 7, 1989, as he was led away after being sentenced to death, he grinned evilly and flashed photographers a devil sign. Earlier, Ramirez had delivered a chilling monologue to the court in which he warned, "I will be avenged."

Conclusion

Without the recently installed fingerprint database, Ramirez might well have remained at liberty much longer at a cost of countless more lives. The inventors of the computer microprocessor could hardly have envisioned that one day it would play such a vital role in fighting crime.

Valerian Trifa

DATE: 1984

LOCATION: Green Lake, Michigan

SIGNIFICANCE: The first American citizen to be deported for war crimes, Archbishop Valerian Trifa had his background mercilessly exposed in a most unexpected and ingenious manner.

In the years after World War II, the United States opened its arms to a vast wave of European immigrants. Most had endured unimaginable levels of hardship and deprivation; a few had not. For this tiny minority migration to America meant more than a chance to carve out a new life; it was their opportunity to bury the past.

One such man was Romanian-born Viorel Trifa. Arriving from Italy, he entered the United States on July 17, 1950. Immigration officials listened with sympathy as this thirty-six-year-old priest of the Romanian Orthodox Church recounted harrowing details of his ordeal in Nazi concentration camps. They granted him immediate residency status. Trifa moved to the Detroit area, where thousands of his fellow Romanians had settled. Ambitious and impatient, he advanced quickly through the church hierarchy, and in 1952 he was consecrated as a bishop. At the same time he adopted the first name of Valerian, a saint who was commemorated every January 21. It was a date that had figured prominently in the life of the man from Romania.

In 1957 Trifa was naturalized as a U.S. citizen. Thereafter, outside the confines of his church, he was virtually unknown and anonymous, and he probably would have remained so had it not been for the unstinting efforts of a few people determined to ensure that justice would be done.

The first uneasy murmurings about Bishop Valerian Trifa had surfaced in 1952. Letters received by the Immigration and Naturalization Service (INS) contained accusations that not only had Trifa been a Nazi sympathizer in wartime Romania, but that he had instigated the Bucharest pogrom of 1941. Why, asked the writers, because the 1948 Displaced Persons Act forbade entry to "any person who advocated or assisted in the persecution of any person because of race, religion or national region," had this murderer been allowed entry into the United States? Surely, having perjured himself on his entry application, he was now subject to deportation? The INS disagreed; it brushed aside every allegation, saying, "This Service cannot interfere in religious matters."

Brutal Pogrom

To describe the events in Bucharest in 1941 as "religious matters" was stretching incredulity to the breaking point. On January 21—St. Valerian's Day—hordes of green-shirted members of the Iron Guard, Romania's virulently pro-Nazi political party, scythed through the Jewish quarter of Bucharest, laying waste to everyone and everything. The madness lasted three days and left almost six thousand dead. An eyewitness wrote, "Perhaps the most horrifying single episode was the 'kosher butchering' of more than 200 Jews in the municipal slaughterhouse. The Greenshirts forced them to undress and led them to the chopping block, where they cut their throats in a horrible parody of the traditional Jewish methods of slaughtering fowl and livestock."

Among the leaders of this massacre was a squat, heavily built man named Viorel Trifa. Later that year, to escape reprisals, Trifa sought and received sanctuary in Germany. Tried in absentia for his part in the atrocities, he was condemned to life imprisonment with hard labor. His whereabouts in the postwar years were shrouded in mystery until his arrival in America in 1950. For two decades, campaigners waged an unceasing struggle to have Trifa's citizenship revoked. Finally, in 1975, the U.S. Department of Justice instituted deportation proceedings against Trifa, alleging that he had concealed material facts in obtaining his citizenship.

By this time, Trifa had been elevated to the rank of archbishop and was in no mood to surrender easily to government pressure. In 1980, stripped of his citizenship, he admitted membership in the Iron Guard, but remained unrepentant. "I am not ashamed of my past at all," he told an interviewer. "For those circumstances in that time I think that

I didn't have any other alternative but to do what I thought to be right for the interests of the Romanian people." And he insisted that claims of high-level connections with Nazis were nonsense. Then, in May 1982, came the evidence that undid him. After searching its archives, the West German government made available to the FBI's Identification Division numerous documents pertaining to Trifa. One such document was a postcard, dated June 14, 1942, allegedly authored by Trifa and addressed to Heinrich Himmler, head of the Nazi SS. Trifa emphatically denied writing the card.

Mindful of Bonn's request that the card not be damaged or defaced, the FBI fingerprint department turned to its latest weapon. With laser technology, a sample can be examined for prints without being dusted or treated in any other way. The argon laser is turned on, and the sample passes under the viewing area. By looking through a protective filtering lens, the examiner can detect latent fingerprints. In this case, the postcard yielded a clear impression of a left thumbprint. When compared to Trifa's, it was identical. The disgraced archbishop's forty-year-old lie was over.

On August 13, 1984, Valerian Trifa left the United States for good and settled in Portugal. In a disquieting interview given shortly afterward, he was characteristically unapologetic. "I am a man who happened to get put in a moment of history when some people wanted to make a point," he said. "The point was to revive the Holocaust. But all this talk by the Jews about the Holocaust is going to backfire." Quite what Trifa meant by this was never made clear. Nor did he get much chance to explain. After suffering a heart attack, he was admitted to a hospital in Cascais on the Lisbon coast, and on January 28, 1987, he died.

Conclusion

Of the millions of Jews slaughtered in World War II, 425,000 were Romanian. Those who survived the horror of Bucharest in 1941 wanted only one thing: justice. Few could have imagined that they would find it in a high-tech beam of light.

Stella Nickell

DATE: 1986

LOCATION: Seattle, Washington

SIGNIFICANCE: Product tampering is one of the most odious of all crimes. Here it was used to mask a callous murder for profit.

As she was getting ready for work on June 11, 1986, Sue Snow, a forty-year-old bank manager in the Seattle suburb of Auburn, collapsed in her bathroom. Paramedics summoned to the scene found Mrs. Snow semiconscious, unresponsive, and gasping for breath. Later that morning, in hospital, she died. Doctors were puzzled. The symptoms suggested either an aneurysm or drug overdose, but neither seemed probable because there was no evidence of internal bleeding, and Mrs. Snow was the kind of woman who restricted her drug use to the occasional painkiller. Coincidentally, earlier that morning she had taken two Extra-Strength Excedrin capsules to combat a persistent headache.

During the autopsy, an assistant detected a faint odor of bitter almonds emanating from the body—a telltale sign of recently ingested cyanide. A laboratory test came back positive. Family members continued to insist that Sue Snow would never have poisoned herself. Yet somehow she had swallowed cyanide, which left only the Excedrin capsules. Might they have been tainted? A follow-up test confirmed that the capsules did indeed contain cyanide.

On June 16, the Food and Drug Administration (FDA) published the lot number of the contaminated capsules, and the manufacturer, Bristol-Myers, cabled stores across the country to take all bottles of Extra-Strength Excedrin off their shelves. Meanwhile, Seattle police found two other bottles of cyanide-laced Excedrin, one in Auburn and the other in Kent, an adjoining suburb.

Already control of the investigation had been ceded to the FBI.[1] Initial suspicions that the killer might be a political terrorist or a disgruntled Bristol-Myers employee soon faded when no one called to either take credit or issue demands. Then, on June 17, a forty-two-year-old

---

1. Product tampering was made a federal crime after the 1982 Tylenol outrage that left seven Chicago-area residents dead. That crime remains unsolved.

widow named Stella Nickell telephoned the police with a strange story. Just twelve days earlier, she said, her husband, Bruce, fifty-two, had died suddenly after taking Extra-Strength Excedrin capsules. She wondered if there could be any connection between his death and that of Sue Snow.

Second Victim

Investigators feared the worst. Although Bruce Nickell's autopsy had recorded the cause of death as emphysema and he had already been buried, an exhumation was unnecessary because the deceased had volunteered to be an organ donor. Consequently, a sample of his blood had been preserved. When that sample was tested, it, too, revealed the presence of cyanide. Even before the results were known, police officers had already recovered two bottles of contaminated capsules from the Nickells' home.

While agents labored in vain to establish a connection between Bruce Nickell, a heavy-equipment operator, and banker Sue Snow, a chemist at the FBI Laboratory found something peculiar: all of the tainted capsules recovered so far contained particles of an algicide used in home fish tanks. He even came up with the brand name—Algae Destroyer. Whoever spiked the capsules, it appeared, had mixed the cyanide in a container used previously to crush algicide pellets.

Another, more sinister, oddity emerged. The FDA had examined more than 740,000 capsules sold throughout the Pacific Northwest and Alaska, yet only five bottles were contaminated, and of those bottles, two were found in Stella Nickell's home. Had she purchased both bottles at the same time, then it could have been ascribed to ill luck, but her claim to have bought them on different days and in different stores defied every known law of probability. The odds against such random misfortune were astronomical.

Immediately, Stella Nickell came under closer scrutiny. As a grandmother with two daughters, she seemed an unlikely killer; neighbors said that she and Bruce had been happy together. It was the same at the Seattle airport, where Stella worked as a security guard. Fellow employees described her as cheerful and hardworking. Bruce's death had devastated her, they said, she had been inconsolable. But the FBI had found out something else—Stella Nickell had a fish tank in her home.

It might have been coincidence, of course, but allied to doubts raised by the bottle episode, this latest revelation catapulted Stella

Nickell into the position of prime suspect. Agents visited local pet stores, anxious to know if anyone recalled a middle-aged woman buying Algae Destroyer. It was a long shot, but on August 25 they got lucky. When shown a photo layout, a store clerk in Kent had no hesitation in identifying Nickell as having bought the algicide from him. He remembered her distinctly because a little bell attached to her purse had jingled as she walked around the store.

As agents delved more deeply into her background, slowly the real Stella Nickell began to emerge. Between 1968 and 1971, while living in California, she had been convicted of check fraud, forgery, and child abuse. Since then she had steered clear of the law but had not managed to steer clear of debt; the Nickells were permanently broke. At the time of Bruce's death, the bank was moving to foreclose on their home. Earlier, they had come perilously close to bankruptcy. And yet, despite this mountain of debt, in the past year Stella had somehow managed to find the money for extra insurance coverage on Bruce's life. As a state employee, he was already insured for $31,000, with an additional $105,000 should death result from an accident. Stella had topped up that sum with another $40,000 of coverage. In total, she stood to receive $176,000 if Bruce's death was judged to be accidental.

All this was well and good, except that the doctor who signed Bruce Nickell's death certificate recorded emphysema as the cause, despite several phone calls from an anxious-sounding Stella, eager to learn if he could possibly have been mistaken in his findings. There was good reason for her concern. Under the provisions of the insurance policies, had Bruce Nickell died from cyanide during a product-tampering scare, then his death would have been ruled accidental and Stella would have pocketed the entire $176,000.

Hostile Interview

On November 18, more than five months after Sue Snow's death, Nickell was brought in for questioning. She first denied ever buying Algae Destroyer, then scoffed at suggestions that she had purchased additional insurance on her husband's life. This second denial—so foolish and unnecessary—immediately branded her as a liar. As the questions became more penetrating she brought the interview to a tearful conclusion, angrily refusing to submit to a polygraph test. Inexplicably, four days later she changed her mind. When asked by the examiner if she had laced the capsules with cyanide, she replied "No." The polygraph needle jumped wildly. So did Nickell. Furious at the outcome, she refused to say another word without benefit of counsel.

For several more weeks the case languished. Then Stella's own daughter, twenty-seven-year-old Cindy Hamilton, contacted the police. Although estranged from her mother, she had felt an understandable loyalty when initially interviewed; now she wanted to clear her conscience. Her mother, she said, had often talked of killing Bruce, even the possibility of hiring a hit man. At other times, she had mentioned cyanide. Although certain that Cindy was speaking the truth, prosecutors knew that any competent defense lawyer would portray this testimony as the product of a disaffected daughter out to gain revenge. While they sought how best to combat this likelihood, Cindy herself provided the answer.

She mentioned that her mother had researched the effects of cyanide at various libraries. Immediately agents visited every local library. In her hometown of Auburn, an overdue notice came to light for a book that Stella Nickell had borrowed and never returned titled *Human Poisoning*. Armed with her card number, an agent searched the aisles for every book that Stella had borrowed. Inside a volume on toxic plants called *Deadly Harvest,* he found her number stamped twice on the checkout slip—both dates before Bruce's death. Acting on a hunch, he sent the book to the FBI Laboratory.

There, fingerprint experts brought the case home to Nickell. Eighty-four of her prints were lifted from the pages of the book, most from the section dealing with cyanide.

On December 9, 1987, Stella Nickell was indicted on five counts of murder and product tampering. Her trial began the following April. Prosecutors painted a damning portrait of a psychopath prepared to murder complete strangers in order to profit from the death of her own husband. She was convicted on May 9, 1988, and sentenced to ninety years of imprisonment.

As often happens in high-profile cases, Nickell has accumulated a large number of supporters who believe her to be innocent. Some claim to have new evidence; others argue that the prosecution case was tainted, with accusations that Cindy's testimony had been driven by a craving for the reward money. In 2001, Nickell herself appeared on the TV program *48 Hours,* defiant as always. "I am not guilty and I won't quit fighting until I prove it," she said. Thus far, her arguments and those of her supporters have fallen on barren ground. In 2001, a court decided that there was insufficient new evidence to grant a new trial.

Conclusion

Despite these recent developments, any objective appraisal of the evidence suggests that Nickell was justly convicted. For this she had no one to blame but herself. Had she been willing to settle for the original $31,000, the circumstances of her husband's death would have gone undetected. Frustration and greed ruined the perfect murder.

FORENSIC ANTHROPOLOGY

Within the human body are 206 bones. In the average male they weigh twelve pounds; in the average female, ten pounds. Together, they form a remarkable and, to the trained eye, informative framework of the body they once supported. They can show how the person lived; any debilitating illnesses the person had, such as rickets or polio; healed fractures; whether the person was right- or left-handed; and even possible clues as to occupation (for instance, waitresses show signs of their arm strength in their bones; their strong side is more developed than the other side).

Several basic questions arise with the discovery of any skeletal remains:

1. What was the age of the person at the time of death? This is a complex and often vexatious subject (see the George Shotton case on page 152).

2. What was the sex of the person? The clearest indicators are the skull and the pelvis. The male pelvis is narrow and steep, whereas the female pelvis is much broader and shallower (a divergence further accentuated by childbirth). Skull variation is most noticeable in the supraorbital ridge and the nuchal crest, both of which are larger in the male.

3. What was the person's race? Using variation in the shape of the eye sockets and the nose, forensic anthropologists can categorize people into one of three racial groups: Mongoloid (Asian), Negroid (African), and Caucasoid (European). In Negroids and Mongoloids, the ridge of the nose is often broad in relation to its height; in Caucasoids, it is narrower.

4. What was the person's height? When a corpse is intact, measurement presents little difficulty, but in cases of dismemberment, when a body or skeleton is incomplete, a measurement can be calculated because of a

fairly standard relationship between the length of the limbs and the total height of the body. This relationship was first noted by Dr. Mildred Trotter, a professor of anatomy at Washington University in St Louis. After World War II, Trotter worked for the U.S. Army in Hawaii, helping repatriate the remains of servicemen killed in action. By studying the long bones of hundreds of servicemen, Trotter devised a formula that was generally accurate within plus or minus three centimeters. For a male Caucasoid the formula is:

$$\text{Length of femur} \times 2.38 + 61.41 \text{ cm} = \text{height}$$

$$\text{Length of tibia} \times 2.52 + 78.62 \text{ cm} = \text{height}$$

$$\text{Length of fibula} \times 2.68 + 71.78 \text{ cm} = \text{height}$$

These figures are for dry bones (without cartilage), and measurements must be made using a special osteometric board, not a tape measure. Obviously, the more bones that are available, the greater the accuracy. Additional tables that attempt to indicate the corpse's build (that is, slender, medium, or heavy) are also available. The most practical applications for these calculations have been in passenger identification after airline disasters, but as the following cases demonstrate, bones and the information they can divulge have prompted some of forensic science's greatest moments.

Michel Eyraud

DATE: 1889

LOCATION: Lyons, France

SIGNIFICANCE: This case was a triumph for perhaps the greatest of the early forensic scientists.

Investigating complaints about a foul riverside stench at Millery, a small town in central France, a council worker traced the smell to a canvas sack hidden among some bushes. Hesitantly, for the odor was abominable, he loosened the ties, only to reel back in horror as the rotted remains of a dark-bearded man slid into view. The body was taken to the Lyons city morgue, a rancid barge anchored in the middle of the River Rhône. There, on August 13, 1889, Dr. Paul Bernard unpeeled the oil burlap that shrouded the corpse and began his examination. Severe putrefaction made it difficult to state any assertions with certainty, but he determined that death had been due to strangulation and put the victim's age at thirty-five.

Four days later, the discovery of a splintered wooden trunk that, judging from its smell, had held the dead body added to the mystery. A railway label indicated that it had been shipped from Paris to Lyons-Perrache on July 27.

News of the gruesome find made headlines all across France and reached the desk of Assistant Superintendent Marie-François Goron at the Sûreté in Paris. He went at once to the missing-person files and found that on Saturday, July 27, a man had reported the disappearance of his brother-in-law, Toussaint-Augsent Gouffé, a forty-nine-year-old bailiff and notorious philanderer last seen the previous day. Goron had visited Gouffé's office in the rue Montmartre. The janitor reported hearing a man go upstairs at nine o'clock on Friday evening and had assumed it was Gouffé. But when the man came downstairs and hurried off, he had realized it was a stranger. In Gouffé's office, Goron had found nothing untoward except for a pile of spent matches on the floor. Dismissing the disappearance as most likely another interlude in Gouffé's eclectic sex life, Goron had left.

But now he wasn't so sure. Neither was he convinced when Gouffé's brother-in-law visited the Lyons morgue, glanced briefly at the corpse, and shook his head, saying that the missing man had chestnut hair, while that of the corpse was black.

Wrong Corpse?

Goron's stubbornness took him to Lyons, despite the advice of the examining magistrate, who assured him that the case was almost solved. A cabdriver had told the police of driving three men and a heavy trunk from the Paris train to a spot near Millery. Yet within hours of his arrival, Goron exposed the cabdriver as a publicity-seeking liar, and the investigation was resumed.

When Goron interviewed Dr. Bernard, the physician defiantly brandished a test tube of hair taken from the unknown man, adamant that the corpse could not possibly be that of Gouffé, because the hair color was entirely different. Bombast turned to disbelief as he watched Goron immerse the hair in distilled water, removing the grime and blackened blood, allowing a distinctive chestnut color to emerge. At Goron's behest, the corpse was exhumed from the cemetery of La Guillotiere and delivered to Jean Alexandre Eugéne Lacassagne, professor of forensic medicine at the University of Lyons and a true pioneer of scientific criminal investigation.

With decomposition so far advanced, Lacassagne devoted most of his attention to the bones and hair. He spent days paring off the maggoty flesh until just a skeleton was left. At once his expert eye picked out an irregularity. A deformity of the right knee had affected the bones to which the muscles are attached; any wasting of those muscles would probably result in a limp. A family member confirmed that Gouffé had fallen from a horse in childhood and had limped ever after. As to the cause of death, Lacassagne agreed that it was strangulation—damage to the thyroid cartilage was manifest—but he thought the signs were consistent with manual strangulation rather than Bernard's verdict of garroting with a ligature. Another point of contention was the age of the corpse. Bernard had guessed thirty-five; after examining the teeth, Lacassagne put it at closer to fifty (Gouffé had been forty-nine). Finally, microscopic comparison of Gouffé's hair, taken from his hairbrush, to that of the corpse convinced Lacassagne that this was Gouffé.

Vindicated, but still with a killer to find, Goron returned to Paris. Never a slave to convention, he offered money from his own pocket as an inducement to loosen tongues. It worked. A friend of Gouffé's described seeing him in a bar with an ugly, balding fellow called Michel Eyraud two days before the murder; also present was Eyraud's attractive young mistress, Gabrielle Bompard. Apparently, Eyraud, who had convictions for pimping, was now nowhere to be found.

Second Trunk

In hopes of jogging memories, Goron hired a carpenter to make a copy of the rotten trunk, which was then exhibited publicly. An estimated thirty-five thousand people trooped by, while photographs were circulated around the world. Within days, Goron received a letter from a Frenchman living in London. He told how the previous July, a man accompanied by his daughter, subsequently identified as Eyraud and Bompard, had taken lodgings with him; before returning to Paris they had bought a trunk very like the one shown in the photograph.

Despite massive newspaper coverage, neither Eyraud nor Bompard was heard from until January 1890, when out of the blue, Eyraud wrote Goron a letter from New York. In it he protested his own innocence but viciously turned on Gabrielle, whom he accused of the murder.

A few days later, this bizarre tale took yet another twist, when Mlle. Bompard appeared at Goron's office to divulge an astonishing saga of greed, sex, and murder. During an assignation with Gouffé at her apartment, she had coquettishly coiled a sash from her dressing gown around his neck, then draped it up to an overhead pulley. This was the signal for Eyraud to leap out from behind a curtain and begin hauling the startled bailiff skyward. But the knot had slipped, and Eyraud had had to strangle Gouffé with his powerful hands. The intent had been robbery but they had overlooked several thousand francs hidden behind some papers. The next day they had taken the trunk to Lyons and then had fled to America. According to Mlle. Bompard, when Eyraud announced his intention to reprise their murderous performance, she had returned to Paris.

On May 19, 1890, Eyraud was captured in Havana, Cuba, and extradited to France. As expected, the trial caused a sensation. Few codefendants have attacked each other quite so virulently as Eyraud and Bompard, drowning their defense in a sea of acrimony. The outcome was never in doubt, only the sentence. Gallic courts have traditionally treated attractive female defendants with great leniency, and Bompard was no exception. She received twenty years in jail; Eyraud lost his head to the guillotine.

Conclusion

Had Eyraud and Bompard shipped their trunk to anywhere else in France, they would, in all probability, have escaped detection. But they chose Lyons, and that brought the trunk and its contents under the scrutiny of Alexandre Lacassagne. His identification of Gouffé's corpse, though an outstanding forensic triumph, was just one of many over a long career (he died in 1921). Before anyone else, he recognized the significance of striations on bullets; he was a leader in bloodstain analysis at crime scenes; later, he undertook groundbreaking research into the criminal psyche. It is a body of work without parallel, and it fully entitles Lacassagne to be called the father of scientific criminal investigation.

Adolphe Luetgert

DATE: 1897

LOCATION: Chicago, Illinois

SIGNIFICANCE: This crime rocked America just as science was beginning to gain a foothold in the courtroom.

As the nineteenth century drew to a close, Chicago's population was multiplying at a prodigious rate. For the industrious, the rewards were high, and by 1897, Adolphe Luetgert, a stocky middle-aged German, had carved out a comfortable living as proprietor of a sausage factory on the north side of the city. Unfortunately, success at work did not translate into happiness at home. An inveterate rake, Luetgert spent more time entertaining various mistresses than he did at home. But a reconciliation seemed to be under way when, on May 1, he and his wife Louisa set out for an evening stroll. It was the last time anyone saw the petite Mrs. Luetgert alive.

Three days later, her brother called at the Luetgert house, anxious to learn Louisa's whereabouts. Luetgert shrugged off his concerns, slyly hinting that Louisa had eloped with a lover. Such a suggestion, absurd and entirely out of character, drove angry family members to the police. A search of local sewers and waterways yielded nothing, and on May 15, they turned their attentions to the sausage factory. A nervous watchman led them down to the cellar, home to three large steam vats. One, containing a malodorous reddish-brown liquid, was emptied and its contents strained through three gunny sacks. Among the residue were what appeared to be particles of bone; they also found two gold rings, the larger of which was engraved "L. L."

More ominous material was found around the vat: a twelve-inch strand of hair, half of an upper false tooth, a piece of cloth, some string, a scrap of leather, and a hairpin. In the smokehouse, among the ashes, fragments of charred bone and corset steel strongly suggested that Luetgert had first attempted to boil his wife to nothing in the sausage vat and then had burned what remained. Analysis of the reddish-brown liquid that oozed from the vat revealed traces of hematin, a decomposition product of hemoglobin normally found in blood.

Such grisly revelations stunned the local Polish and German communities, who, sickened by the prospect of their staple diet having been defiled in so dreadful a manner, packed the public gallery to overflowing when Luetgert's trial opened at Cook County Criminal Court on August 30, 1897.

Murder Plans

They heard the story of a venal, vicious brute who beat his wife without remorse, flaunted his affairs in front of her, and repeatedly boasted of plans to get rid of her. These plans appeared to have been put into motion on March 11, when, according to sausage factory employee Frank Ordowsky, Luetgert arranged delivery of 378 pounds of crude potash to the factory. The following month, on April 24, Ordowsky and another employee were told to empty the potash into barrels alongside the vat. Later that day, Luetgert tipped the potash, which "burned like fire," into the vat, filled it with water, and let it simmer for a week.

On the day of Louisa's disappearance, a Saturday, Luetgert opened a valve, allowing steam into the vat. Once the mixture was boiling, he dispatched the watchman on errands that left the factory empty most of the day and that night. All weekend the vat bubbled. On Monday morning, Luetgert scraped up the residue that had boiled onto the floor, burned some of it in the smokehouse, buried yet more, and dumped the rest next to some railroad tracks. Curious employees were told that he was making soap to clean the premises prior to selling them.

This, then, was the prosecution's case. To allay defense doubts that such a concoction would actually dispose of a body, they cooked their own potash soup, into which they then lowered a human cadaver. They got similar results.

The forensic team entrusted with examining the remains was headed by Dr. George Dorsey, a Harvard-trained anthropologist. He and his colleagues from the Field Museum recovered part of a femur, a phalanx (toe bone), a rib, a sesamoid (an extra bone that sometimes forms over tendons, especially near the joints in fingers and toes), a fragment of skull, a metacarpal, and part of a humerus. Dorsey, tetchy and somewhat theatrical on the stand, demonstrated what to him were clear similarities between the charred remains and specimens from the museum, assuring the court that the bone fragments, which barely totaled a few ounces, were definitely those of a human female.

With the benefit of hindsight, modern anthropologists have expressed misgivings about these assertions, disputing whether enough material had been recovered to positively identify it as being human in origin. Many of Dorsey's contemporaries shared this unease. However, any hopes entertained by Luetgert that medical conflict would aid his defense were dashed by the abundance of corroborating evidence. He had no answer for the engraved gold wedding band, or the witnesses who testified that Louisa's knuckles were so swollen from arthritis that slipping the ring from her finger was an impossibility—unless another means had been found.

Convicted of first-degree murder, Luetgert served a life sentence at Joliet State Penitentiary and died from a heart attack in 1911.

Conclusion

Sadly, the criticism that Dorsey endured from other, more circumspect anthropologists persuaded him never to testify again in a trial. Even if his ego failed to survive the mauling, his legacy did; without innovators like Dr. George Dorsey, forensic testimony would never have achieved its modern credibility.

George Shotton

DATE: 1919

LOCATION: Swansea, Wales

SIGNIFICANCE: An incredible tale of deceit, murder, and intrigue baffled investigators for decades.

By the spring of 1920, detectives in South Wales were convinced that glamorous ex–chorus girl Mamie Stuart, twenty-six, missing since the previous November, had been murdered. Even more galling was the belief that they knew who had killed her—George Shotton, a smooth-talking marine surveyor. The only fly in the ointment, as far as police were concerned, was that Mamie Stuart's body was nowhere to be found.

She had married Shotton in 1918, and after much moving around, the couple had settled in a remote cottage overlooking Swansea Bay. Suddenly, just before Christmas 1919, both had vanished. It so happened that at about the same time, a male guest had departed a Swansea hotel, leaving a leather trunk. When by the following March the trunk had still not been claimed, the hotelier called the police. Inside were two dresses and a pair of shoes, all slashed to ribbons; some jewelry; a Bible; a rosary; and a manicure set. There was also a scrap of paper bearing the address of Mamie Stuart's parents. Mr. and Mrs. Stuart identified the contents as belonging to their daughter and announced that they had been trying in vain for months to contact her. Both feared for her safety. They handed over a letter written by Mamie in which she expressed grave doubts about her husband. It read, "If you don't hear from me, please wire to Mrs. Hearn [a friend] and see if she knows anything about me. The man is not all there. I don't think I will live with him much longer. My life is not worth living." Mrs. Hearn confirmed that Mamie had once begged, "If I am ever missing, do your utmost to find me, won't you?"

Soon afterward, a cleaner at the cottage found a leather handbag behind the bedroom dresser. It contained a sugar ration card in Mamie's name and two pounds (eight dollars) in change. By this time, Chief Inspector William Draper of Scotland Yard had discovered what Mamie Stuart never knew—that George Shotton had been married for years. Furthermore, he and his wife and child were living in an isolated house barely a mile from the cottage that he and Mamie had called home.

Shotton admitted knowing Mamie and leaving the trunk at the hotel, but he denied that they had married. He claimed to have last seen her in early December, when they had quarreled over her infidelity and parted. Corroboration for at least one facet of this story came with the discovery of a letter penned by Mamie to an admirer. It read, "My old man seems to know quite a lot . . . but what the eye don't see the heart can't grieve. . . . Am just dying to see you and feel your dear arms around me."

Acting on the belief that Shotton, consumed by jealousy, had killed Mamie and disposed of her body, Draper ordered every inch of the cottage and garden searched. Nothing was found. Police fanned out and scoured the surrounding countryside, but all to no avail. The missing woman's description, circulated throughout Britain, also failed to produce a single worthwhile lead, which meant that the charge sheet when Shotton was arrested on May 29, 1920, read "bigamy" alone. His defense—that someone else had assumed his identity and married Mamie in July 1918—failed to impress the court, and Shotton began a sentence of eighteen months' hard labor.

And there the case of the missing ex–chorus girl might well have ended—but for a strange turn of events forty-one years later.

Four Decades Pass

On Sunday, November 5, 1961, three cavers in a region of the South Wales coast known as Brandy Cove were exploring a disused lead mine when they made a gruesome discovery. Behind some rocks, almost hidden by a thick stone slab, lay a sack containing human bones. Nearby was a black butterfly comb—a tuft of dark brown hair still attached—and two rings, one a gold wedding band.

The remains were taken to the Forensic Science Laboratory at Cardiff, where Home Office pathologists Dr. William James and Dr. John Griffiths reassembled the bones into a complete skeleton. Cause of death was undetermined; it might have been strangulation or stabbing, but there was nothing to say definitely. The body had been sawn into three pieces, and judging from the pelvis, it was female. She had been approximately five feet four inches tall, the same height as Mamie Stuart. Close examination of the skeleton revealed how old the woman had been at death. Until the midtwenties, gristle at the end of the bones, especially the long bones, is soft to allow for growth. Eventually this gristle calcifies and becomes bone—as in this case—a process usually completed by age twenty-five. When the rest of the skeleton has consolidated, the sutures of the cranium begin to fuse together, again at a predictable rate, until completely sealed by age thirty. Here, none of the sutures were closed, which put the woman between the ages of twenty-four and twenty-eight at the time she died.

In sex, height, and age, the skeletal remains matched Mamie Stuart's description. Scraps of clothing and shoes found in the mine shaft came from the 1920s. An elderly lady, a close friend of Mamie's, identified the rings as belonging to the long-lost girl, while hallmarks put the date of manufacture between 1912 and 1918. Conclusive identification came by superimposing a photograph of the skull over a life-size portrait taken of Mamie during her theater days. The result left no room for doubt—after forty-one years, Mamie Stuart's body had been found.

At the coroner's inquest, the court heard a startling tale. Bill

Symons, an eighty-three-year-old ex-postman, recalled an afternoon in 1919 when he had been delivering mail and happened to see George Shotton struggling with a heavy sack outside his cottage. Shotton looked up, saw Symons's blue uniform, and nearly fainted. Gathering his wits quickly, he exclaimed, "God! For a minute I thought you were a policeman." When Symons offered to help carry the sack to Shotton's van, which was parked on the road, Shotton hastily declined and hefted the sack alone. The incident had ended with Shotton driving off in the direction of Brandy Cove.

After consideration of this evidence, the coroner's jury decided that the skeleton was indeed that of Mamie Stuart, that she had been murdered, and that the evidence pointed to George Shotton as the murderer. All of this then raised the inevitable question—where was George Shotton?

Three weeks later, after a search involving Interpol, the ex–marine surveyor was tracked down—to a cemetery in the west of England. He had died penniless and alone at age seventy-eight in a Bristol hospital on April 30, 1958. Forensic science had revealed his secret, but just three years too late to confront him with the evidence.

Conclusion

Nowadays in Britain, no coroner's jury would be allowed to name a suspected murderer; the Criminal Law Act of 1977 withdrew such privileges. Ironically, the last person so named was Lord Lucan, who vanished just after the 1974 killing of his children's nanny. He, too, has never been found.

William Bayly

DATE: 1933

LOCATION: Ruawaro, New Zealand

SIGNIFICANCE: This case concluded with a landmark trial in which the defendant was accused of double murder, even though the body of one of the victims was never found.

On Monday, October 16, 1933, the body of Christobel Lakey was dragged from a pond on the farm that she and her husband, Samuel, worked near Ruawaro, a town on the north island of New Zealand. Her face, a mass of purple welts, provided hideous confirmation that she had been murdered. Sam Lakey was nowhere to be seen, and all of his rifles were missing. An old buggy that had been unused for a decade showed signs of recent use, and its wheels were stained with fresh blood. More blood found in a barn suggested that someone had shot Sam Lakey indoors and then had used the buggy to dispose of his body. But where?

That remained a mystery until October 30, when a policeman probing swampland near the Lakeys' boundary fence felt something metallic. He scraped away several inches of mud to find the missing firearms—two shotguns and a pea rifle, all later identified as belonging to Lakey. Just then, the owner of the adjoining property, William Bayly, stormed up and angrily ordered the searchers off his land. Police were unsurprised by his belligerence. They had already heard stories of bad blood between Bayly and the Lakeys, the result of a dispute over grazing rights. Allegedly, one quarrel had terminated with Christobel shouting, "You murdered Elsie Bayly and your conscience is hurting you. I wouldn't be surprised if we got the same treatment!"[1]

Assured by police that they had no intention of trespassing on his land, Bayly became more cooperative, enough to volunteer that perhaps Sam Lakey, upon finding his wife's body, had panicked and fled, fearing he would be suspected of murder. Curiously, after signing a statement to this effect, Bayly also vanished. Soon afterward, a neighbor came forward, talking of thick smoke belching from Bayly's cowshed on the day that the Lakeys had disappeared. Police went at once to investigate.

Cremated Body

A fine carpet of ash covered the cowshed floor. In one corner stood a large, smoke-blackened oil drum, surrounded by blood. Immediately fear grew that Lakey had been incinerated in the makeshift crematorium, and a full-scale survey of the farmyard seemed to bear this out. Searchers found hundreds of bone fragments, all roasted; two false teeth; some sacking; part of a cigarette lighter; and the stem of a cherrywood pipe, similar to one owned by Lakey. Scrapings of shot lead taken from the oil drum and elsewhere in the shed totaled 28.7 grams, coincidentally the exact weight of a bullet from Lakey's rifle. A shell of this size was also found in a pair of Bayly's dungarees. Draining a

---

1. Some years earlier, Bayly's cousin Elsie Walker had died in suspicious circumstances. Bayly was arrested but was later released due to a lack of evidence.

sheep dip revealed yet more bones together with a tuft of gray hair, some shotgun cartridges, and a cigarette lighter whose white wool wick was identical to wicks found in Mrs. Lakey's workbasket.

Although investigators were convinced that Bayly was a double murderer, to refute any future defense claims that Lakey had killed his wife and then fled, they wanted confirmation that these body parts had belonged to Sam Lakey.

New Zealand's premier pathologist, Dr. P. P. Lynch, was asked to conduct a thorough examination of the remains, a mammoth undertaking. He first dealt with the hair—it was definitely human in origin. As for the individual bones, he identified the atlas, which is the top bone of the vertebral column; a portion of heelbone; and numerous parts of the vault, or top of the skull. In order to test the theory that Bayly had cremated Sam Lakey's body in the large oil drum, Lynch burned the carcass of a 150-pound calf in a similar container. The fire was started at quarter past four one afternoon. By the next morning, the carcass was reduced to a bucketful of ash, much the same amount as was found in the cowshed.

Before Lynch could complete his work, Bayly came out of hiding and gave himself up. His trial in Auckland for double murder lasted almost a month. Dozens of witnesses testified on behalf of the prosecution, but none did so more effectively than Lynch. He produced no fewer than 250 exhibits—jars, boxes, bottles, and vials—all containing what he claimed were the residual parts of Sam Lakey. All the defense could counter with was a weak claim of conspiracy by the police to manufacture evidence against the accused.

On June 23, 1934, history was made. Even without Sam Lakey's corpse to provide evidence against him, Bayly was convicted of both murders, and a month later he was hanged at Mount Eden Jail.

Conclusion

Even today, juries are reluctant to return murder convictions without the presence of a body. That a New Zealand jury was willing to take this unprecedented step so many years ago speaks volumes about the quality and quantity of forensic evidence made available by the prosecution.

John Wayne Gacy

DATE: 1978

LOCATION: Des Plaines, Illinois

SIGNIFICANCE: The sheer enormity of Gacy's crimes made it virtually impossible to identify all of his victims. Discovering their names required the most painstaking analysis.

When police called at the Des Plaines, Illinois, home of contractor John Wayne Gacy on December 13, 1978, there was nothing to suggest that America's worst serial killer was about to be unmasked. They were there in response to a routine missing-person report; fifteen-year-old Robert Piest had applied for a job with Gacy and had not been seen since. Gacy, thirty-six, managed to deflect their inquiries, but when a follow-up background check revealed a past conviction for sexually molesting a minor, officers hastened back to the house on West Summerdale Avenue. Virtually unbidden, Gacy showed them a trapdoor that led into a crawl space under the house.

When raised, the trapdoor revealed a hideous sight: bodies everywhere, all strangled and raped. Gacy coolly admitted that twenty-seven bodies were buried either beneath his house or in the garage (there were actually twenty-eight—he had forgotten one). When he ran out of space around the house, Gacy dumped another five bodies, including Robert Piest's, into a nearby river.

A vicious homosexual, Gacy had cruised Chicago's Bughouse Square district, trolling for young males, enticing them into his black Oldsmobile with offers of marijuana. Those who succumbed to his savage sexual demands were released, bleeding and battered; those who didn't were chloroformed into insensibility, raped, and then killed. Gacy tapped into another source of victims by offering five-dollar-per-hour jobs with his contracting business. Few lived to draw a paycheck.

Identifying the remains provided investigators with a huge logistical problem. First, there was the similarity of the victims—all males between ages fourteen and the midtwenties. Next, many parents, unwilling to accept that their sons might have drifted into a lifestyle that they found repugnant, were reluctant to extend any assistance. It was left to dental charts, X-rays, and fingerprint records to provide the names that the authorities so desperately sought. By the end of

January 1979, only ten had been identified. Frustrated by such dilatory progress, investigators requested the assistance of Clyde Snow, an Oklahoma anthropologist with an encyclopedic knowledge of bones.

Because of Gacy's haphazard burial techniques—bodies were piled on top of bodies—Snow first had to ensure that bones from different skeletons had not been confused by the excavation crew. Next, he compiled a thirty-five-point reference chart of each head for comparison with various missing-person reports.

One of these yielded the name of David Talsma, a nineteen-year-old ex-Marine reported missing in the Chicago area on December 9, 1977. As a boy, Talsma had broken his left arm. Snow recalled that one of the skeletons—the seventeenth disinterred at Gacy's house—had shown signs of similar distress. At five feet eleven inches, the skeleton was the same height as Talsma, and hospital records confirmed that Talsma had sustained a head injury, as had this body. A Holmesian element was introduced with Snow's final finding: The left arm was several millimeters longer than the right. Together with beveling of the left scapula, this strongly suggested that this victim had been left-handed—as Talsma had been. One more successful identification had been completed, but there were still many more to go.

All through 1979 the team labored, but by the year's end nine bodies had still not been identified. With frustration mounting, Snow decided to expand the search parameters.

Forensic Sculpture

The art of reconstructing a face from a skull is not a recent development; it dates back to 1895, when Swiss-born anatomist Wilhelm His began creating full-size models that often bore remarkable similarities to his subjects. On one occasion, he was given a skull thought to be that of Johann Sebastian Bach. By carefully reconstructing the face, His confirmed that the skull did indeed belong to the great composer. Obviously, eye and hair color remained areas of guesswork, but His found that the bone conformation provided him with a clear idea of what the face would look like.

One of the foremost modern practitioners of facial reconstructions, Betty Pat Gatliff, frequently collaborated with Snow. Given seven of the unidentified skulls to work with, she first set about establishing skin thickness by gluing pencil eraser heads to strategic points on the skull. Then modeling clay was applied in accordance with certain carefully delineated measurements. These gave Gatliff an idea of what the mouth and cheeks would look like, but not the nose, which always presents special difficulties. Because cartilage decomposes quickly after death, leaving the characteristic gaping hole, only by rearranging the slivers of bone that remained could Gatliff make a reasonable estimate of what the nose's shape might have been. The final touches were cosmetic: prosthetic eyes, and in cases in which hair had been found with the bodies, the addition of an appropriately colored wig.

When photographs of the reconstructions were released to the media, the response was disappointing. Not a single definite identification resulted, although independent phone calls from two Chicago sisters hinted that one of the reconstructions bore a very strong resemblance to their missing brother. Tragically, neither girl would proceed due to the parents' unwillingness to become involved. Right to the end, this perceived lack of parental concern plagued the investigation, with the result that nine of Gacy's victims remain unidentified.

Conclusion

On March 13, 1980, Gacy was convicted of thirty-three murders, the highest tally ever in the United States, and he was sentenced to death. His fourteen-year battle against that fate ended on May 10, 1994, when he was executed by lethal injection.

Josef Mengele

DATE: 1985

LOCATION: Embú, Brazil

SIGNIFICANCE: Invariably when forensic science attempts to identify a corpse, its efforts are directed toward the victim of a homicide or accident. On this occasion the roles were oddly reversed: The victims were known, and they numbered in the millions. But was their killer this handful of bones recovered from a Brazilian hillside?

When Russian soldiers entered the gates of Auschwitz on the afternoon of January 27, 1945, it was the outside world's first glimpse of high-tech genocide. Hundreds of corpses littered the ground, all that was left of the estimated 1.25 million people who died in this remote corner of Poland, sacrificed on the altar of Aryan racial purity. Only the devastation remained; the architects of this madness had already disappeared into the transit camp that was postwar Europe, among them the man who would become the most hunted criminal on earth.

The search for Dr. Josef Mengele was slow in getting under way. At first his name meant nothing to the Allies. And he was lucky. After a brief period in American hands—they had no idea who he was—he finally made his way to Argentina in 1949 and there lived a life of virtual anonymity. Ironically, it took the capture of another Nazi, Adolf Eichmann, in 1961, to awaken interest in Mengele. At long last the mad doctor, the so-called Angel of Death, who had personally consigned four hundred thousand people to their doom, took his place alongside the other great ogres of history. Details of hideous experiments on children helped fuel the image of a maniac hell-bent on engineering the perfect race from a test tube.

As interest grew, so did the myths. Books and movies hinted at jungle hideouts where an increasingly deranged Mengele coordinated efforts to resurrect the Third Reich, funded by bulging Swiss bank accounts and assorted Latin dictators. His family insisted that he was dead. His pursuers believed otherwise. International indifference, widespread and entrenched, finally gave way in 1985, when Washington announced that there would soon be official efforts to bring Mengele to justice.

But it was too late. With rumors of Mengele's imminent capture gaining strength, a German couple living in Brazil, Wolfram and Lisolette Bossert, brought all speculation to an abrupt halt by leading police to the village of Embú and the grave of a man buried under the name of Wolfgang Gerhard. This, they said, was the final resting place of Josef Mengele, dead from drowning in 1979 at age sixty-seven.

Exhumation

Exactly forty-one years to the day after Allied troops landed on the beaches of Normandy, Dr. José Antonio de Mello, assistant director of the local forensic department, opened up the coffin. Its contents were taken away for inspection by scientists from the United States, Germany, and an independent team representing the Simon Wiesenthal Center in Vienna, who had pursued Mengele for decades. Never before had such a formidable team been assembled to discover the identity of a corpse.

Conformation of the pelvis—narrow and steep—suggested that the skeleton was male; bones on the right side, being markedly longer than those on the left, pointed to someone who had been right-handed. The eye sockets and nose were Caucasian. Several factors helped establish the subject's age at the time of death. Because tooth abrasion occurs at a constant rate throughout a lifetime, allied to the amount of adjacent bone loss this can give an approximate guide to the age. Here it suggested someone between sixty and seventy years old, very close to what Mengele's age would have been. Microscopic examination of the femur's blood-carrying canals (the more numerous and the more fragmented they are, the older the individual) also hinted at someone in his late sixties.

Because Mengele's SS file contained only the skimpiest physical details—white male, 174 centimeters (5 feet 8.5 inches) tall, with a head circumference of 57 centimeters (22.4 inches)—scientists had to tread carefully. A hand-drawn dental chart from 1938 showed twelve fillings but failed to note their type or exact location. Also, there was no mention of Mengele's distinctive gap-toothed smile. Despite the paucity of medical records, some compelling conclusions were drawn. Measurement of the femur, tibia, and humerus bones suggested a height of 173.5 centimeters, within half a centimeter of Mengele's known height, and an X-ray of the teeth, compared in minute detail with the 1938 chart, showed an unusually wide incisor canal between the halves of the upper palate that must have produced a gap-toothed smile.

Richard Helmer, a forensic anthropologist from West Germany, provided the clinching evidence. A champion of video superimposition, in which a photograph is placed over an image of a skull to determine whether the two are the same person, Helmer first marked more than thirty identifying features on the skull. Then, with two high-resolution video cameras mounted on tracks, he began shooting. Using known photographs of Mengele and a photograph of the skull, Helmer matched the images exactly, allowing the forensic team to state that, within reasonable scientific certainty, the bones at Embú were those of Josef Mengele.

Not everyone was pleased. Some victims and conspiracy buffs, particularly journalists and writers who had a vested interest in Mengele's continued existence, all expressed doubts. The Angel of Death, they said, had tricked the world again and was still in hiding, dreaming up his mad schemes for world domination.

Conclusion

Finally, in 1992, the myth was well and truly laid to rest. Genetic material from the bones found in Brazil was compared with samples from Mengele's living relatives in Germany. Using a technique that the evil doctor could never have imagined, DNA analysis confirmed the match. The hunt for the century's most wanted criminal was indeed over at last.

John List

DATE: 1986

LOCATION: Westfield, New Jersey

SIGNIFICANCE: Bones reveal details not only about the victims of crimes but occasionally about the perpetrators as well.

For almost a month, the large, run-down house owned by John List in Westfield, New Jersey, had stood deserted, and yet each night the lights still blazed brightly. His neighbors on Hillside Avenue were puzzled. They knew the forty-six-year-old insurance salesman as a tightwad, hardly the kind to run up unnecessarily high electricity bills. Repeated phone calls to the police finally resulted in action on the night of December 7, 1971, when two officers went to investigate. After knocking and receiving no answer, they gained entrance through a window. Inside it was icy cold. Their flashlight tour of the rambling eighteen-room mansion came to a sudden halt in a large room at the rear of the house. Neither man could believe his eyes. Spread out neatly on the floor were four sleeping bags. On top of each, just as tidily arranged, was a dead body.

John List's wife, Helen, had been shot through the head, as had been her three teenage children. Upstairs, her eighty-five-year-old mother-in-law, Alma List, had also been executed. Only John List was missing. Five addressed envelopes were taped to a filing cabinet. The letters they contained provided chilling insight into the mind of a man who had reached the end of his mental tether. List, a pillar of the local Lutheran church, readily conceded the wrongfulness of his actions, but he rationalized that they had been necessary to spare his family the ignominy of a life on welfare. Financial foolhardiness had led him to the brink of bankruptcy, and rather than let his sons face ridicule at the hands of friends, he had decided to kill them. The same went for his wife and mother. Because his daughter's interest in acting and the occult would compromise her spiritual development, she, too, had to die.

Pathologists estimated that death had occurred a month earlier. By tracing family members' movements, detectives were able to narrow this down to one day, November 9. It appeared as though Helen List had been shot in the morning, then Alma was shot just minutes later. That afternoon, when the children returned, List eliminated them one by one. In all, he fired twenty-three rounds of ammunition from two weapons, a .22 caliber Colt revolver and a 1918 German nine-millimeter Steyr semiautomatic pistol. As the investigation swung into action, detectives were confident that they would soon be able to clear this case from their books; after all, they already had a written confession. Even the fact that List had disappeared was seen as only a temporary setback; he would be easy enough to find, they reasoned.

The first clue to his whereabouts came just forty-eight hours after the bodies were found, when List's car was discovered in the long-term parking lot at New York's JFK Airport. Police also discovered that he was not completely without funds; bank records showed that on the day of the killings, List had withdrawn $2,100 in cash from his mother's account. But as days drifted into weeks and then months, the expected capture did not materialize. A coast-to-coast wanted-poster campaign also failed to produce any kind of worthwhile lead. Because of his fluency in German, circulars were distributed in West Germany and German-speaking regions of South Africa. Again, there was nothing. More frustration followed when investigators targeted areas where the Lutheran church was prominent. By now, the months had meandered into years, and still there was no sign of the missing man—seemingly, John List had vanished off the face of the earth.

Resolute Detective

When Bernard Tracey first joined the Westfield police force, List had already been a fugitive for two years. Nevertheless, the rookie cop acquired an almost obsessional interest in the unsolved case, little thinking that it would haunt him for almost sixteen years. Promotion to the rank of detective allowed him the opportunity to add his own input to the search for the elusive killer. But every avenue led nowhere. By early 1986, almost at his wit's end, Tracey contacted the supermarket tabloids in an effort to resurrect interest in the case. On February 17, the *Weekly World News* ran an account of the List family murders and included a photograph of the fugitive. One reader in Aurora, Colorado, saw the photograph and gasped. Wanda Flanery was amazed by the uncanny resemblance between List and her neighbor's husband, Bob Clark. Unable to contain herself, she confronted Mrs. Clark with the article. Delores Clark, who knew nothing of her husband's background, thought the idea preposterous and said so, defending her husband in such a forthright and heartfelt manner that Wanda wound up apologizing in embarrassment for having raised the subject. Only later would Wanda Flanery learn that she was not mistaken. List had moved to the Denver area within two weeks of the murders and had applied for a Social Security number in the name of Robert P. Clark. Under this alias he had created an entirely new life for himself.

List never heard about this brush with disaster. Delores had found the idea so ridiculous that she didn't even mention it to him. The way she figured it, her husband might be hopeless with money, but he certainly wasn't any killer! She was half right: List's fiscal ineptitude had not deserted him. Not long after this incident, with debts piling up around his ears, he was forced to take an accounting job back in Richmond, Virginia. In 1988, Delores joined him.

TV Program

That same year, Tracey played his final card. He contacted the makers of *America's Most Wanted,* a syndicated TV show that ran true-crime stories and had a good record of tracking down wanted fugitives. (The program was mandatory viewing in the home of Richmond accountant Bob Clark!) But the show's producers gave Tracey a cool reception, explaining that they rarely featured cases more than eight years old; any older and the trail was, they felt, too cold for them to be of any practical assistance.

But Tracey was indefatigable. A year later he tried again, and this time the program's producers said yes. The most significant aspect of their involvement was the decision to hire forensic sculptor Frank A. Bender of Philadelphia, who had undertaken similar work for the U.S. Marshals, to produce a bust of John List as he might look eighteen years after the crime.

Because appearance is so heavily influenced by diet and exercise, Dr. Richard Walter of the Michigan Department of Corrections was asked to prepare a psychological profile of List with special emphasis on his likely lifestyle. Walter felt that List's religious upbringing would proscribe any attempt to alter his appearance by plastic surgery. That same background would also manifest itself in a "meat and potatoes" type of person. Because nothing suggested a fondness for exercise, such an unrelieved diet would very likely produce sagging jowls as List aged, giving him the appearance of being somewhat older than he actually was.

Assistance came in the form of a revolutionary new computer program that simulates the effects of aging—sags, wrinkles, receding hairline. With this software, FBI technical specialist Gene O'Donnell was able to scan the last known photograph of John List, at age forty-six, into the computer and then generate a picture of his likely current appearance.

Using all of these tools, Bender produced a bust of the fugitive that showed heavy jowls, a receding hairline, and gray hair. Because List had not been the outdoor type, Bender made the skin coloring pale. As a final touch, he added thick-lensed spectacles, much thicker than those worn by List in 1971.

On May 21, 1989, *America's Most Wanted* broadcast its segment on John List. After reenacting the crime, it showed the bust in profile and full face. Of the more than 250 calls received, one came from Colorado. The anonymous caller said that List was currently living in Richmond, Virginia, under the alias Robert P. Clark. (The caller turned out to be a relative of Wanda Flanery, who had watched the show in Colorado and had again noticed the similarity to her ex-neighbor.) After several days of working their way through all the tips received, investigators finally interviewed "Bob Clark" on June 1, 1989.

Loud protestations of mistaken identity fizzled upon production of his fingerprints, and the sixty-four-year-old man who had been on the run for almost two decades was returned to New Jersey to stand trial. On April 12, 1990, he was jailed for life.

Conclusion

The use of computer-enhanced imagery is growing fast. In 1994, London police enlisted the facilities of *Vogue* magazine to help with an unknown dead woman whose face was too battered to permit publication of a photograph. Using equipment designed to manipulate photo images of models so that every blemish is removed, experts were able to create a photograph of how the woman had looked when alive. Within hours she was successfully identified.

ODONTOLOGY

Of all the tissues in the human body, nothing tends to outlast teeth after death, a characteristic which makes them ideally suited as a means of identification. Indeed, in the aftermath of serious fires, teeth are often the only means of identifying scorched remains. It is frequently claimed that no two people have identical teeth. However, one must keep in mind that, unlike fingerprints which remain unchanged from birth, dentition achieves its uniqueness through use and wear. For successful identification, both ante- and postmortem records must be available. From such data, it is often possible to make an identification from a single tooth.

It is estimated that almost two hundred different tooth-charting methods exist throughout the world. The American approach, called the universal system, allocates a different number to each of the thirty-two adult teeth, beginning with the upper right third molar (1) and continuing around the mouth to the lower right third molar (32). Information is recorded about the five visible surfaces of each tooth, thus making it possible to complete a dental grid or odontogram, unique to the individual. Naturally, odontological features that can identify remains may also help identify criminals if bite mark evidence is left at the crime scene.

A bite mark is not necessarily an accurate representation of the teeth. Much depends on the mechanics of jaw movement and use of the tongue. Inside the mouth, the lower jaw (mandible) is movable and usually delivers the most biting force. The upper jaw (maxilla) is stationary, holding and stretching the skin, but when skin is ripped or torn, the upper teeth are involved more deeply.

For forensic purposes, bite marks fall into two categories:

1. Human-on-human, in which the skin of one or more participants in an assault is penetrated. (This may not always be the victim. Quite often the

victim bites the attacker in self-defense.) The primary value of the bite mark is to identify (or exclude) a suspected assailant.

2. Bite marks on food, such as cheese, chocolate, or fruit. This may establish the presence of a person at a particular crime scene.

Of the two, bite marks on humans pose the greatest problem, as they may alter with the passage of time. For this reason, bite marks are routinely photographed over a set period of hours or days, so a permanent record is made. (Ultraviolet light can reveal bite marks even months after they have been inflicted.) If there is no penetration, the underlying bruising may take up to four hours to develop in the living, and is clearly visible for up to thirty-six hours. In a dead victim, they may take twelve to twenty-four hours to become visible. Sometimes it is possible to make a silicone rubber cast if the bite is deep enough. Before this can be done, swab specimens must be taken from the site, as residual saliva can often be detected and used for blood typing or DNA analysis.

Teeth are also useful in determining the age of a corpse, particularly someone young. Because the dental tissue growth of four microns per day is registered by striations on the tooth, it is possible to estimate the age of a young person within twenty days on either side. Age can also be assessed by X-rays of teeth in the jaws and dental eruptions into the mouth, but because these developments are affected by stimuli such as diet, race, and environment, they are not useful after the early teens.

After age twenty-five, senile changes are associated with the teeth; they wear down on the biting surfaces, gums recede, the pulp chamber becomes smaller, the roots are resorbed, and the tips of the roots become translucent. Only by careful observation of all these factors can the odontologist hazard an age. Even then, accuracy is generally no better than about forty-two months either side of the true age.

John Webster

DATE: 1849

LOCATION: Boston, Massachusetts

SIGNIFICANCE: This case involved the first American convicted of murder without the victim's body being either wholly present or definitely identified.

Shortly before Thanksgiving 1849, prominent Bostonian Dr. George Parkman visited the laboratory of Dr. John Webster, a professor of chemistry and mineralogy at Harvard Medical College. The visit was far from social. For some time, Parkman had been hounding Webster for repayment of a $438 loan and had publicly threatened to ruin the professor's career unless the obligation was met. That night Parkman failed to return home. Two days later, on Sunday, November 25, his wife had posters distributed throughout Boston offering a reward for information leading to her husband's whereabouts. Within days, Harvard College increased the reward to three thousand dollars. But there was still no sign of Dr. Parkman.

Webster told Parkman's family that he had repaid his debt on the Friday in question and hinted that Parkman might have been waylaid by thieves on his way home. After all, he was carrying a considerable sum of money.

Someone less than convinced with this story was Ephraim Littlefield, a college janitor. He recalled that on the day of Parkman's disappearance, the door to Webster's laboratory had been locked, and yet a wall adjacent to the lab's assay oven was red-hot. When Littlefield mentioned this to Webster, the latter blanched and blurted out that he had been conducting experiments. The next day, in a rare expression of seasonal goodwill, Webster presented the inquisitive janitor with a Thanksgiving turkey. This only served to make Littlefield even more suspicious. With his wife standing guard, he chiseled through the brick wall using a crowbar and peered into the gloom. In his own words, he described seeing "the pelvis of a man and two parts of a leg," adding with commendable understatement, "I knew that it was no place for these things!"

Littlefield quickly summoned the police. Their search of Webster's laboratory uncovered even more evidence of wrongdoing. In a wooden chest lay a human thorax, the sternum and ribs that constitute the upper torso. And in the oven that Webster used to heat various chemicals,

SPECIAL

NOTICE!

GEO. PARKMAN, M. D.,

A well known, and highly respected citizen of BOSTON, left his House in WALNUT STREET, to meet an engagement of business, on Friday last, November 23d, between 12 and 1 o'clock, P. M., and was seen in the Southerly part of the City, in and near Washington Street, in conversation with some persons, at about 5 o'clock of the afternoon, of the same day.

Any person who can give information relative to him, that may lead to his discovery, is earnestly requested to communicate the same immediately to the City Marshall, for which he shall be liberally rewarded.

BOSTON, Nov. 25th, 1849.

From the Congress Printing House,(Farwell & Co.) 32 Congress St.

Reward notice issued in 1849 to publicize the disappearance of Boston physician George Parkman. (Library of Congress)

they found what would become the most crucial evidence of all—a set of false teeth. When Webster was arrested, crowds surrounded his house, bombarding his terrified family members with insults and stones. Elsewhere, the shock waves reverberated far beyond Harvard and Boston, as all across America people devoured newspaper accounts of the grisly investigation.

More than 150 bones or fragments were recovered from the laboratory. The task of identifying them was delegated to a team of Webster's university colleagues. Comprising some of the finest scientific minds in America, the panel concluded that the remains were human and belonged to a male, approximately five feet ten inches tall, between ages fifty and sixty—very similar to Parkman, who was sixty years old and had stood an inch under six feet.

Webster's claim that the body parts were from a cadaver used by medical students was refuted by the team's inability to detect any traces of embalming fluid. They were divided on whether the dissection showed evidence of anatomical skill—Webster, a chemist, had not dissected a body since leaving school—but agreed that the level of decomposition coincided with the length of time Parkman had been missing. Also, the remains came from someone very hirsute, as was Parkman. Although unable to determine the precise cause of death, the panel entertained little doubt that these were the mortal remains of George Parkman.

Suicide Attempt

In custody, Webster attempted suicide with strychnine, but he survived to face a charge of murder. His trial caused such a sensation that the public gallery was operated on a ten-minute shift system to accommodate all the spectators. More than sixty thousand people trooped through the courtroom for a glimpse of the notorious defendant.

The testimony of dentist Dr. Nathan Keep was crucial. A close friend of both the victim and the accused, he described how, four years previously, Parkman had come to him for dentures. Because of Parkman's unusually prominent jaw, Keep was obliged to fashion a special cast. Nevertheless, when Parkman tried the dentures he complained of soreness and chafing, so Keep filed them down. On the witness stand, Keep pointed out those file marks to the jury. He also demonstrated how the dentures fit his mold exactly.

For the defense, Dr. William Morton—a pioneer of ether use in dentistry—openly derided Keep's assertion that this set of dentures, and this set alone, fit the cast he had made. To prove his point, Morton produced another set that also fit neatly into Keep's cast.

But public opinion weighed heavily against Webster, as did the final direction of Judge Lemuel Shaw to the jury, in which he stated that the corpus delicti, or actual crime itself, need only be proved "beyond a reasonable doubt." This was a landmark decision. A team of America's most distinguished anthropologists had established only that the bones were those of an elderly man, falling short of positively identifying them as belonging to Parkman. Without the dental testimony, it is doubtful that Shaw would have dared to utter such a directive.

At the end of an eleven-day hearing, Webster was found guilty; he was hanged before an unruly mob on August 30, 1850. Just before his execution, he made a full confession. Just as investigators had thought, the two had met in Webster's laboratory in the college basement. An argument broke out, with Parkman shouting: "I got you your professorship, and I'll get you out of it." Webster terminated the dispute by bludgeoning Parkman with a piece of wood and dissecting the body in a sink. The larger parts went into a vault used for storing cadavers; the rest were incinerated in his assay oven.

Conclusion

Despite the questionable identification—prosecutors nowadays would shun such skimpy proof—the Webster case remains a forensic milestone. It demonstrated for the first time the strength of multitasking in scientific crime investigation. Each expert contributed what he could, and on this occasion the sum of the whole was enough to justify the end result.

Harry Dobkin

DATE: 1942

LOCATION: London, England

SIGNIFICANCE: This case involves one of the earliest examples of odontological identification.

By the summer of 1942, German air raids had reduced so much of London to rubble that just trying to keep up with the business of removing debris was a major logistical nightmare. On July 17, a cleanup crew visited a bombed-out Baptist chapel in Kennington. Down in the crypt, workers began clearing the wreckage. Beneath a stone slab they found a corpse, little more than a skeleton with only a few shreds of skin intact. At first, everything suggested that this was just another victim of the Luftwaffe's nightly air raids. Yet as they lifted the body from its grave, the skull just came away. It had been severed from the body. When pathologist Dr. Keith Simpson examined the remains in the Southwark mortuary the next day, he determined that the victim was female and that she had been dead between a year and eighteen months.

A close inspection confirmed early suspicions that this was a case of willful murder. Not only the head but both legs had been severed, and then the corpse was doused with slaked lime—calcium hydroxide—in an ill-judged attempt to hasten decay. It was ill judged because, contrary to popular belief, slaked lime tends to preserve organic matter, not destroy it. Enough of the larynx was left for Simpson to detect a small blood clot on the upper horn of the right wing of the voice box, almost certainly the result of strangulation.

Judging from the sutures of the skull vault—the brow plates were completely fused and fusion was under way between the top plates—Simpson estimated the woman's age at between forty and fifty. Measurement of the bones suggested someone approximately five feet tall. The hair that remained was dark brown, just turning to gray. Simpson deduced that at some time she would have received medical attention for a fibroid tumor that had enlarged the uterus, and she had obviously undergone extensive dental treatment. Teeth in the upper jaw contained several fillings and showed the marks of a dental plate. Simpson knew that here lay his best chance of a definite identification.

Missing Persons

Acting on Simpson's report, the police studied a list of missing persons. The difficulties of compiling such a list in wartime are obvious, but one name stood out—Rachel Dobkin, the forty-seven-year-old wife of fire warden Harry Dobkin, who was employed by the insurance company next to the Baptist chapel to keep an eye out for fires. For almost twenty years, he and his estranged wife had been locked in a bitter dispute over nonpayment of child support, a conflict that had earned the recalcitrant Dobkin several prison terms. His wife, five feet one inch tall, had been reported missing on April 11, 1941. The next day, her identification card, ration card, and rent book were found at a post office in Guildford, Surrey. Her sister told the police that Rachel had gone to collect arrears of support from her husband. The couple had left a cafe in Dalston at 6:30 P.M., and Rachel had not been seen since.

Mrs. Dobkin's sister also provided the name of Rachel's dentist,

Dr. Barnett Kopkin of North London. He had kept precise records of Mrs. Dobkin's treatment from 1934 to 1941 and was able to sketch a chart of her upper jaw. In conformation and number of teeth, location of fillings, and marks made by a dental plate, Dr. Kopkin's diagram matched a photograph of the corpse's mouth exactly. Also, in April 1941, he had extracted two teeth from Mrs. Dobkin's lower left jaw, an operation that left portions of the roots intact. An X-ray of this jaw, examined by a colleague of Simpson's at Guy's Hospital, Sir William Kelsey Fry, showed those same roots.

When Rachel Dobkin's doctor, Marie Watson of the Mildmay Mission Hospital, confirmed that the missing woman had had a uterine fibroid tumor, that should have been enough to satisfy any court, but Simpson, ever the perfectionist, removed any lingering doubts by superimposing a photograph of Rachel Dobkin on the skull, the technique Professor John Glaister had pioneered in the 1930s (see the Buck Ruxton case on page 226).

At his trial, Harry Dobkin's lawyers adopted a two-pronged defense. First, they attempted to dismiss Simpson's findings as being informed by speculation and not fact; then they suggested that if the victim was Mrs. Dobkin, in all probability she had been killed in an air raid. Neither approach worked. Dobkin was convicted and sentenced to death; he was hanged on January 27, 1943.

Conclusion

Simpson's testimony in this trial brought him national prominence. During his long and distinguished career, his meticulous attention to detail and impeccable fairness made him one of the world's foremost pathologists, a man whose opinion was respected and requested around the globe.

Gordon Hay

DATE: 1967

LOCATION: Biggar, Scotland

SIGNIFICANCE: At the outset of this case, no one realized that a single bite mark would make legal history.

An all-night search for fifteen-year-old schoolgirl Linda Peacock ended at 6:45 A.M. on August 7, 1967, with the discovery of her lifeless body in a cemetery in Biggar, a small country town southwest of

Edinburgh. She had fought furiously for her life until she was clubbed senseless and then strangled with a rope. Although she had not been raped, her bra and blouse were disarranged, and on the right breast was an oval-shaped bruise. Crime scene photographs included no less than fifteen photos of the bruise, which appeared to be a bite mark. Scotland's premier forensic odontologist, Dr. Warren Harvey, analyzed the bruise and confirmed that it was, indeed, a bite mark and that one of the killer's teeth was uncommonly jagged.

On the night of Linda's disappearance, witnesses had seen a couple talking by the churchyard gates at around 10:00 P.M. A woman who lived nearby reported hearing a girl scream about twenty minutes later. Although no one could say for sure, the girl matched Linda's description, and she appeared to know the young man she was talking to. At that time, the population of Biggar totaled about two thousand, yet within a week almost twice that number had been interviewed and eliminated as police officers widened their search for the killer. Ultimately the investigation focused on a local detention center for young offenders. Twenty-nine inmates were asked to provide dental impressions. Models made from the casts were then compared with transparencies of the wound.

Purely by observation, Dr. Harvey reduced to five the number of impressions that could not yet be eliminated from suspicion. These five inmates were recalled for further dental impressions. At this stage, Harvey consulted Professor Keith Simpson, the Home Office pathologist, who later said that in more than thirty years, he had never seen a better-defined bite mark. Together, the two experts concentrated their attention on the jagged tooth.

They soon whittled the second batch of impressions down to just one. Gordon Hay, seventeen, had been arrested the previous year for breaking into a factory. He was surly and truculent; his detention had been littered with frequent clashes with authority, and yet he willingly submitted to a further set of impressions being made of his teeth. Harvey and Simpson studied the sharp-edged, clear-cut pits like small craters on the tips of the upper and lower right canines. They were caused by a rare disorder called hypocalcination. The upper pit was larger than the lower and matched exactly the bite mark on the victim's breast. Harvey, in no doubt that they had found the killer, described Hay's teeth as "absolutely unique."

Fantastic Odds

During the course of the investigation, Harvey had examined 342 teenage soldiers. Of those checked, only two had pits, one had a pit and hypocalcination, and none had two pits. Extrapolation of these findings throughout the general population led Harvey to conclude that even if seventeen people could be found with two pitted canines and hypocalcination, then it would still be virtually impossible for any of them to have left the exact mark found on Linda Peacock's breast, so individual was that single tooth.

Hay's defense in the face of this forensic blitzkrieg was one of alibi; he swore that at the time of Linda Peacock's death, he was in the detention dormitory. Not so, said another inmate. He claimed that Hay had entered the dormitory much later that night, breathless, disheveled, and with mud on the knees of his jeans. Yet another boy described Hay meeting Linda Peacock at a local fair the day before her murder. Afterward, as she walked away, Hay announced his intention to have sex with her. (Apparently, local girls frequently dated boys from the detention center. On this occasion, it sounded as though a brief encounter had ended in tragedy when Linda refused Hay's sexual demands. Security at the center was later tightened to prevent any more nocturnal assignations.)

At Hay's trial, which began at Edinburgh High Court on February 26, 1968, his lawyers tried everything to get the dental evidence ruled inadmissible. When that failed, they countered with dental experts of their own. For the first time, a Scottish jury had to consider conflicting bite mark testimony. Deciding that the prosecution witnesses had made their case, they found Gordon Hay guilty of murder. Because of his age, he was sentenced "to be detained during Her Majesty's pleasure."

Conclusion

An appeal claiming the nonadmissibility of dental identification was argued ferociously but failed to overturn the verdict. With their decision to uphold the earlier judgment, the court sent a clear message that bite mark testimony was here to stay.

Wayne Boden

DATE: 1968

LOCATION: Montreal, Canada

SIGNIFICANCE: The first North American case in which forensic bite mark evidence played a crucial role.

By the middle of January 1970, the worst suspicions of the Montreal police had been confirmed—a serial killer was on the loose in their city. Four times in the preceding eighteen months, young women had died in gruesomely similar circumstances. The first was a popular twenty-one-year-old teacher named Norma Vaillancourt. Her nude and strangled body was found in her apartment on July 23, 1968. Although the killer had gnawed repeatedly at her bare breasts, leaving deep bite marks, there were no signs of resistance; indeed, the medical examiner described the victim's facial expression as serene, smiling almost. This led investigators to wonder if, during consensual sex, the killer had "playfully" strangled the unsuspecting young woman, who had lapsed into unconsciousness without any awareness of what was happening. It was learned that Norma enjoyed an eclectic and varied social life with many boyfriends, none serious. All were checked and cleared.

The following year the body of Shirley Audette, twenty, was found lying on an apartment patio in West Montreal. Although fully clothed, she, too, had been raped and strangled and bitten on the breasts. Again there was that baffling lack of resistance, suggesting that she had known her killer. This murder provided investigators with their first clue. In recent days, Shirley had shared concerns with her friends about an unnamed man, someone who scared her. Unfortunately, she had not elaborated on his identity.

Unlike most serial killers who tend to murder at random, the man branded the Vampire Killer invariably became well acquainted with his victims. Marielle Archambault was a case in point. Other employees at the jewelry store where this attractive twenty-year-old worked remembered her leaving the store at closing time on November 23, 1969, with a handsome, flashily dressed young man. She had seemed ecstatic in his company. The next morning Marielle's desecrated body was found on the living room floor of her apartment. Unlike the previous victims, she had fought furiously for her life, but ugly teeth marks on her breasts provided grim confirmation that the Vampire Killer had claimed victim number three. A photograph of a young man was found at the apartment. Marielle's co-workers identified him as the man who had called for her the previous afternoon, but a sketch made from the photo and printed in the newspapers failed to produce any leads.

The last Montreal victim was twenty-four-year-old Jean Wray. When her boyfriend knocked on the door of her apartment at 8:15 P.M. on January 16, 1970, there was no reply. Puzzled because they had a dinner date, the boyfriend went for a beer in a nearby bar, then returned. This time the door was unlocked. Jean was lying on a sofa, totally nude, and with those trademark bite marks all over her body. Once again there was no sign of a struggle and her face looked eerily tranquil. Police believed that when the boyfriend had first called, the murderer was still in the apartment.

Cross-Country Killer

For more than a year the Vampire Killer lay dormant. His next eruption occurred half a continent away from Montreal. When Elizabeth Pourteous, a high school teacher in Calgary, failed to report for work on May 18, 1971, her apartment manager was asked to check if everything was all right. He found Elizabeth on the bedroom floor. The scene was depressingly familiar: raped, bra torn open to reveal savage bite marks on her breasts and neck.

The luck that had so conspicuously eluded the Montreal investigation fell right into the hands of the Calgary detectives charged with finding Elizabeth's killer. Two fellow teachers had seen her in a blue Mercedes with a young man the previous evening. They remembered that the car had a bumper sticker advertising beef. Another friend reported that Elizabeth had been very excited about the new man in her life, a well-dressed fellow named Bill, who in every detail matched the description of the Montreal killer.

Next day the blue Mercedes was spotted close to the victim's apartment. A police stakeout produced almost immediate results. Within half an hour a young man was arrested as he approached the car. There could be no doubt that he was the person in the photograph found in Marielle Archambault's apartment. He gave his name as Wayne Boden, formerly of Montreal, a twenty-three-year-old ex-model who had moved to Calgary a year earlier. He admitted being with Elizabeth on the previous evening, and that he had left a cuff link at her apartment—it was found during the autopsy, embedded in

Elizabeth's back—but he insisted that she was still alive when he left the apartment.

An examination of Boden's underwear revealed seminal stains and pubic hairs that matched those from the body of Elizabeth Pourteous. But the decisive evidence was provided by a Calgary orthodontist, Dr. Gordon Swann, who identified the bite marks on Elizabeth Pourteous's breasts as having twenty-nine points of similarity with Boden's teeth.

Boden was sentenced to life imprisonment. He later confessed to three of the four Montreal killings—he always resolutely denied murdering Norma Vaillancourt—and received three more life terms. Officially, the murder of Miss Vaillancourt remains unsolved. In 1984, Boden hit the headlines again, when, on a "humanitarian" day pass from the maximum-security Laval Correctional Centre in Quebec, he escaped from a Montreal hotel after asking to use the bathroom. The panic over the Vampire Killer's liberty lasted less than twenty-four hours. The next day he was recaptured in a Montreal bar. He remained behind bars for the rest of his life. His sentence ended on March 27, 2006, when he died of natural causes at the Kingston Penitentiary in Ontario after a short illness.

Conclusion

In an unusual move, after being convicted, Boden agreed to discuss his case with Swann. During the course of this meeting, Boden expressed astonishment and fascination with the amount of information that Swann was able to glean from the bite mark evidence. "I didn't think you'd appreciate it that much," he told the surprised orthodontist. Then, in a moment of unexpected frankness, he added darkly, "I realize I have a problem."

Theodore Bundy

DATE: 1978

LOCATION: Tallahassee, Florida

SIGNIFICANCE: Ted Bundy's dazzling smile not only attracted victims, it also sent him to the electric chair.

In the early hours of January 15, 1978, someone broke into the Chi Omega sorority house at Florida State University (FSU) in Tallahassee and went on a blood-drenched rampage. At its conclusion, two students, Lisa Levy and Margaret Bowman, lay dead; two others were grievously injured. A man wearing a stocking cap and carrying a wooden club was seen fleeing from the building. Less than ninety minutes later, another student was attacked just a few blocks away, but fortunately she survived. Amid the carnage at the sorority, one vital piece of evidence emerged. Lisa Levy had sustained bite marks on her left buttock. Before the marks were photographed, an officer had the presence of mind to include a yellow ruler in the photo to give a sense of scale. This would later prove to be critical.

One month later, a man using the name Chris Hagen was arrested in Pensacola for driving a stolen vehicle. When a check of records revealed that "Hagen" was none other than Ted Bundy, wanted felon and suspected serial killer, immediately suspicion grew that he might be the architect of the Tallahassee bloodbath.

Between 1969 and 1975, a tidal wave of sex killings had swept from California through the Pacific Northwest and into Utah and Colorado. Numbering in the dozens, all of the victims were strikingly similar—female, young, attractive, generally with long hair parted in the middle. Some were found dumped in deserted areas, whereas others simply vanished, never to be seen again. The logistical problems of interstate murder investigations are many, but as the various law enforcement agencies compared notes and suspects, one name kept cropping up—Ted Bundy, a handsome young Seattle law student, gregarious and much traveled. Wherever Bundy was, women died. But finding the kind of evidence needed to convince a jury required more than speculation.

And then, on November 8, 1974, eighteen-year-old Carol DaRonch was duped into entering a Volkswagen outside a Salt Lake City shopping mall by a stranger claiming to be a police officer. When the man produced some handcuffs and attempted to bludgeon her, Carol DaRonch fought her way out of the car. Despite this close call, the killer continued to find victims until August 16, 1975, when a Salt Lake City police officer arrested a Volkswagen driver who had been acting suspiciously. Inside the car he found a crowbar and handcuffs. The driver turned out to be Ted Bundy. Amid rumors that he was the slayer of countless women, Bundy was identified by Carol DaRonch as her abductor and was sentenced to one to fifteen years of imprisonment.

One year later, in June 1977, following extradition to Colorado on a murder charge, Bundy escaped. After eight days, he was recaptured. Incredibly, on December 30, 1977, Bundy escaped again. Two weeks and two thousand miles later, the Chi Omega massacre occurred in Florida.

"Organized" Killer

Curiously enough, despite the appalling nature of the attack, the Tallahassee crime scene displayed an eerie neatness and was completely devoid of fingerprints. According to the FBI's Behavioral Science Unit, serial killers tend to fall into two categories: organized and disorganized. The latter kill without any consideration of the consequences, often littering the crime scene with clues. Organized killers, as the term implies, are far more calculating and go to quite extraordinary lengths to conceal their involvement. Unsurprisingly, this category of killer normally poses the greater problem for crime investigators. Yet in this instance, such precautions actually assisted the prosecution.

When William Gunter, the Leon County sheriff's department crime scene specialist, visited Bundy's Tallahassee apartment soon after his arrest, he routinely dusted the place for fingerprints. After a few minutes, he was puzzled. No matter where he searched—on closet doors, shelves, bedposts, handles, even an overhead lightbulb—he could not find a single print. As he testified later, "The room had been wiped clean." Such precautions, typical of the organized killer, only heightened suspicion against Bundy. After all, what kind of person lives in a fingerprint-free environment?

But this was conjecture. The prosecution needed solid evidence linking Bundy to the Chi Omega killings. Their best hope lay with the bite marks on Lisa Levy's left buttock. When Bundy refused to provide impressions of his teeth, a search warrant was issued, authorizing detectives to obtain the examples by force if necessary. Realizing the hopelessness of his situation, Bundy acquiesced. Coral Gables dentist Dr. Richard Souviron began by taking frontal color photographs of Bundy's uneven upper and lower teeth and gums. Then, using a mirror, he obtained a reverse-image photograph of the inside surface of the teeth. Next, Bundy was told to bite into a malleable compound and remain motionless for a few minutes. After a while the compound set, forming a permanent mold. Souviron finally took individual wax impressions of each tooth. By pouring sculpting material into these molds, he was able to make precise stone casts of Bundy's teeth.

After much legal wrangling (Bundy had originally accepted a plea bargain, only to change his mind), his trial for the FSU killings opened in Miami on June 25, 1979.

Deadly Testimony

Brimming with arrogance, Bundy, the onetime law student, insisted on leading his own defense. He tackled the state's witnesses with a cross-examination technique that vacillated between the amateurish and the abysmal—until Dr. Richard Souviron assumed the stand. At this juncture, Bundy threw one of his long-suffering lawyers into the fray and slunk off into the background. This was testimony that could kill him and he knew it.

Souviron was questioned at length about the photograph of the bite marks on Lisa Levy's body and the yellow ruler that appeared in the photo. The defense tried to capitalize on the fact that the original ruler had been lost, until Souviron pointed out that it obviously had existed and that he had no reason to doubt its accuracy.

Using an enlarged photograph of the bite mark on Lisa Levy's skin and a similarly sized photo of Bundy's teeth with the lips retracted, Souviron showed how an acetate overlay of Bundy's front teeth fitted exactly atop the photo of the bite marks. Asked by the prosecution if he thought, within a reasonable degree of certainty, that Bundy's teeth had made the bite marks, Souviron replied, "Yes, sir." For the first time, there was actual physical evidence linking Bundy to a murder victim.

Defense attorney Ed Harvey quickly sought to undermine the setback. "Analyzing bite marks is part art and part science, isn't it?" he asked the dentist.

"I think that's a fair statement."

"Your conclusions are really a matter of opinion. Is that correct?"

Souviron agreed that they were, but nonetheless the damage had been done. Confirmation came from Dr. Lowell Levine, chief consultant in forensic dentistry to the New York City Medical Examiner, who told the court that dental identification was nothing new, having been admitted into testimony as far back as the nineteenth century.

On July 23, 1978, Bundy was found guilty on all charges. More than a decade later, on January 24, 1989, he went to his death in Florida's electric chair. Although the exact number of his victims will never be known, just hours before his execution he hinted that it was somewhere between forty and fifty.

Conclusion

It is impossible to overemphasize the importance of odontological evidence in the trial of Ted Bundy. Even with such testimony, prosecutors knew that their case was far from ironclad; without it, they were almost certainly facing acquittal.

Carmine Calabro

DATE: 1979

LOCATION: New York, New York

SIGNIFICANCE: In this case, astute odontological analysis overcame a freakish crime scene faux pas.

On the evening of October 12, 1979, the nude body of a young woman was found on the roof of the Pelham Parkway housing project in the Bronx. Francine Elveson, twenty-six, a tiny woman less than five feet tall and weighing only eighty pounds, shared an apartment in the building with her parents and had not been seen since that morning when she had left for her teaching job at a nearby day-care center.

Her injuries were horrendous: She had been beaten, strangled with the strap of her purse, and then mutilated in the worst way imaginable. Across her chest, scrawled in ballpoint pen, was an obscene message from the killer to the police, challenging them to track him down. In his frenzy he had launched a ferocious biting attack on the insides of the victim's thighs.

During the autopsy, a single black pubic hair was discovered on the corpse. Obviously not from the victim—it was Negroid in origin—the hair was assumed to have been shed by the attacker, and therefore it was considered of critical importance. Only much later would its true significance become known. With the investigation crystallizing into a search for an unknown black male, it was necessary to more closely examine the bite marks.

Examination of bite marks involves carefully photographing them for future reference; if they are sufficiently prominent, cast impressions are made. These are then compared to the teeth of a possible suspect using a variety of equipment that may include infrared and ultraviolet photography, electron microscopy, even computer analysis. The horror aroused by Francine Elveson's death meant that when detectives requested that local residents provide prints of their own teeth, they met with almost universal acceptance. But none matched the marks found on the body.

As the months passed without a tangible lead, police asked the FBI's Behavioral Science Unit to provide a profile of the likely killer. Their response was essentially as follows: a male, twenty-five to thirty-five, poorly educated, probably living in the building where the attack occurred, either alone or with a single parent, and suffering from a psychosis so acute that it would have required treatment in a mental institution. It was also thought likely that given the saturation police coverage, the killer or someone in his family had already been interviewed. One final point: The killer was almost certainly white, not black. (See the Richard Chase case on page 192.)

Case Review

This last point turned the investigation upside down. A review of previous suspects was undertaken, and gradually, from out of the pack, one candidate emerged. At the time of the murder, police had interviewed a middle-aged man living on the fourth floor of the building—the same as Francine Elveson. The man's son, thirty-two-year-old Carmine Calabro, had a history of mental instability, but according to the father, he was undergoing treatment at a nearby hospital. The police had taken the statement at face value and neglected to double-check it. That oversight was now rectified.

Hospital records corroborated the father's story—Calabro had been a patient for one year—but a study of security at the unit found it so slack that he could easily have slipped out, committed the murder, and returned without anyone being the wiser. Also, at the time of the murder, Calabro had been wearing a cast on his arm, which raised the possibility that he had used his cast to render the victim unconscious.

Since leaving the hospital, Calabro, a high school dropout, had worked as a stagehand until being fired. He had no objection to providing a dental print. When forensic odontologists Dr. Lowell Levine, Dr. Homer Campbell, and Dr. Richard Souviron examined that print, they were in no doubt that Calabro was the man who had sunk his teeth into Francine Elveson's thighs.

On December 20, 1980, more than a year after the murder, Calabro was arrested. He was later jailed for life.

Conclusion

So what about the black pubic hair found on Francine Elveson's body? The solution to that puzzle highlights the ease with which criminal investigations can be compromised if forensic procedures are not carried out to the letter. Apparently, the bag used to transport her body to the medical examiner had previously been occupied by a black male murder victim. Afterward, it should have been thoroughly cleaned, but on this occasion some detritus was overlooked. The unexplained black hair had come from that earlier murder victim—not from Elveson's body at all.

PSYCHOLOGICAL PROFILING

Behind every criminal act is the criminal mind. It sounds like a simple enough concept, and in seeking to probe the motivations of that mind (and thereby improve their "clear-up rate"), detectives have increasingly turned for assistance to what is popularly known as psychological profiling. No other forensic technique has aroused such controversy, particularly in Britain. At the heart of psychological profiling is the belief that criminals leave psychological clues at the scene of the crime. By carefully sifting through these clues, skilled interpreters build a character sketch of the culprit. Common sense, observation, and geographical considerations play as big a role in this process as psychology, for only by studying all facets of the crime can the profiler hope to succeed.

The concept of psychological profiling is far from new (a rudimentary attempt was made to profile the kidnapper of Charles Lindbergh Jr. in 1932), but only recently has it been embraced by the law enforcement community. Time and cost constraints dictate that it be used for only the most serious of crimes. As an adjunct to traditional investigative techniques, its uses are manifold, but a few early and spectacular successes promoted a dangerous overconfidence in its effectiveness. British police, in particular, have felt the sting, culminating in the spectacular abandonment of a 1994 murder trial after it was discovered that the prosecution had no evidence, just a profile.

For all the controversy, psychological profiling is clearly here to stay. Knowing the type of criminal one is seeking has many obvious advantages, but with the advantages comes danger. Maintaining objectivity on a case-by-case basis is not easy. By its very nature, profiling tends to be retrospective, and although solutions to crimes of the present may be suggested by crimes of the past, the shrewd profiler has to remain aware that humankind's capacity for evil innovation is seemingly infinite.

George Metesky

DATE: 1940

LOCATION: New York, New York

SIGNIFICANCE: Profiling's first major success came with the capture of a bomber who had terrorized New York City for years.

One of the most extraordinary and longest-running one-man crime waves ever began on November 16, 1940, with the discovery of an unexploded bomb on a window ledge at the Consolidated Edison building in Manhattan. Wrapped around it, a hand-printed note read, "CON EDISON CROOKS—THIS IS FOR YOU." Police attributed the amateurish device to someone with a grudge against the company that supplies New York with its electricity and filed the case away.

In September 1941, another unexploded bomb was found lying on 19th Street; its alarm-clock fusing mechanism had not been wound. Three months later, following Pearl Harbor, an odd letter was received at police headquarters in Manhattan. Postmarked Westchester County, it read, "I WILL MAKE NO MORE BOMB UNITS FOR THE DURATION OF THE WAR—MY PATRIOTIC FEELINGS HAVE MADE ME DECIDE THIS—I WILL BRING THE CON EDISON TO JUSTICE—THEY WILL PAY FOR THEIR DASTARDLY DEEDS. . . . F.P." As before, the letter was printed in ink on plain white bond paper.

Between 1941 and 1946, some sixteen similar letters were received by newspapers, hotels, and department stores, and Con Ed itself. Then, on March 29, 1950, a third bomb turned up on the lower level of Grand Central Station. This, too, failed to explode, but it showed ominous signs of improvement in its construction. The next bomb did work, devastating a phone booth at the New York Public Library on April 24, 1950. More followed; some exploded, some did not. Miraculously, there were no injuries, despite one being planted inside a theater seat. Eager to allay public concern, the authorities downplayed the bomber's activities, a reticence that aroused fury in the letter writer, who threatened to step up his attacks on theaters.

Between 1951 and 1952, four more bombs exploded, making it impossible to conceal the activities of the man branded the Mad Bomber. Another four devices exploded in 1953. The following year, more were detonated, including one stuffed inside a cinema seat, which injured four people, two seriously. In 1955, there were six bombs; two of those that failed to explode were found in cinema seats. The letters became longer, more vitriolic. One to the *Herald Tribune* boasted, "SO FAR 54 BOMBS PLACED—4 TELEPHONE CALLS MADE—THESE BOMBINGS WILL CONTINUE UNTIL CON EDISON IS BROUGHT TO JUSTICE." Again it was signed "F.P."

Public Outrage

The next bomb, on December 2, 1956, was the most powerful thus far. It blew apart several seats in Brooklyn's Paramount Theater, injuring six people, and it triggered a major campaign to catch the Mad Bomber. To this end, Inspector Howard E. Finney enlisted the aid of Dr. James A. Brussel, a psychiatrist in private practice who also served as New York State's assistant commissioner for mental hygiene. Brussel was noted for his ability to analyze the facts of a case and arrive at some remarkable deductions. Finney, who had a degree in forensic psychiatry, provided Brussel with copies of every letter the bomber had written, together with all other salient information.

Brussel's conclusions were as follows: "It's a man. Paranoiac. He's middle-aged, forty to fifty years old, introvert. Well proportioned in build. He's single, a loner, perhaps living with an older female relative. He is very neat, tidy, and clean-shaven. Good education, but of foreign extraction. Skilled mechanic, neat with tools. Not interested in women. He's a Slav. Religious. Might flare up violently at work when criticized. Possible motive: discharge or reprimand. Feels superior to his critics. Resentment keeps growing. His letters are posted from Westchester, and he wouldn't be stupid enough to post them from where he lives. He probably mails the letter between his home and New York City. One of the biggest concentrations of Poles is in Bridgeport, Connecticut, and to get from there to New York you have to pass through Westchester. He has had a bad disease—possibly heart trouble." As an afterthought he added, "When you catch him, he'll be wearing a double-breasted suit—buttoned."

When Brussel suggested that publicizing his theories might flush the attention-hungry bomber out, a New York paper eagerly took up the challenge, printing an open letter that urged the bomber to give himself up and offering a forum for his grievances. Three letters, all excoriating Con Ed, arrived in quick succession. One read, "I DID NOT GET A SINGLE PENNY FOR A LIFETIME OF MISERY & SUFFERING—JUST ABUSE." Another actually included the date of the incident that had so enraged the author—September 5, 1931.

A search of the Con Ed archives hit pay dirt. In a file that had been sealed since 1937, office assistant Alice Kelly found a letter from a disgruntled former employee that included several of the same stilted phrases used by the bomber. It came from a George Metesky, who had been knocked down by a boiler explosion while working at a Con Ed plant on September 5, 1931. Although he had complained of pains and headaches, doctors were unable to find any physical injuries. Metesky was given sick pay and other insurance benefits for twelve months; then he was fired. When he tried to sue Con Ed, he was advised that his action was too late—all compensation claims had to be filed within two years of the date of the injury. For years, Metesky had brooded over this perceived injustice, and then he began planting bombs.

Profile Comes to Life

Detectives who arrested Metesky at his home in Waterbury on January 21, 1957, discovered a bomb-making workshop in the garage. They also found a well-proportioned man, fifty-four, of Polish extraction, unmarried, living in a house with two older sisters, and wearing a double-breasted suit—buttoned! It was a stunning success for psychological profiling.

Metesky freely admitted to being the Mad Bomber, revealing that the initials "F.P." stood for Fair Play. On April 18, 1957, beaming delightedly, he was found unfit to plead and was committed to Matteawan Hospital for the criminally insane. In 1973, he was pronounced cured and released. All charges against him were dropped.

Brussel was modest about his astonishing triumph, attributing it to simple deductive reasoning, experience, and playing the percentages. His reasoning went as follows: Because paranoia takes a considerable time to develop—often as much as ten years—and the first bomb had been planted in 1940, he felt the man's illness would have started around 1930, making him middle-aged in 1956. Why a paranoid? Because they are the champion grudge holders; feel themselves intellectually superior; and are neat, obsessive, and tidy to a fault—hence the meticulous printing and the double-breasted suit.

Although the wording in the notes suggested an educated mind, they contained no slang—native New Yorkers say "Con Ed" instead of "the Con Edison"—and read as if they had been translated into English. Also the phrase "dastardly deeds" hinted at a foreigner. Why a Slav? Because historically, bombs have been favored in Central Europe. Well proportioned? In a broad-based study, German psychiatrist

Ernst Kretschmer had demonstrated a correlation between a person's build, personality, and any mental illness that might develop. Kretschmer found that 85 percent of paranoiacs have an athletic build. Here, Brussel was going with the averages. Why single? Because of all the neatly printed capitals, only the *W* was curved, like two *U*s joined together—similar to a woman's breasts. This hinted at a sexual problem, quite possibly someone who had never married.

In fact, just about the only thing Brussel got wrong was the heart disease, but even there he didn't miss by much. The Mad Bomber actually suffered from a tubercular lung.

Conclusion

The uncanny accuracy of Brussel's profile did much to raise public awareness of the need to expand the parameters of detection. Detectives thereafter had to probe not only the physical evidence but the psychological ramifications as well. Crime solving would never be the same again.

Richard Chase

DATE: 1978

LOCATION: Sacramento, California

SIGNIFICANCE: FBI profilers painted a devastatingly accurate portrait of a bewildered and malevolent psyche.

On Monday, January 23, 1978, the city of Sacramento was jolted by one of the most brutal murders in living memory, when David Wallin came home in the early evening to find his twenty-two-year-old wife, Theresa, butchered in their bedroom. Wallin ran screaming from the house, incoherent with terror, unable to speak of the horror he had just witnessed. When homicide detectives arrived, they fully shared his revulsion; no one could recall such carnage. Apart from the appalling injuries inflicted on Theresa Wallin, a crushed yogurt container by the side of the body showed signs of having been used by the murderer to drink the blood of his victim. Clearly this crime was far beyond the scope of most homicides, and for that reason a call for assistance went out to the FBI's Behavioral Science Unit at Quantico, Virginia.

Robert Ressler, a longtime FBI instructor in the art of criminal profiling, joined his California-based colleague Russ Vorpagel, and together the two men compiled a provisional profile of the killer, based on the skimpy facts already known. It read as follows:

> White male, aged 25–27; thin, undernourished appearance. Residence will be extremely slovenly and unkempt and evidence of the crime will be found at the residence. History of mental illness, and will have been involved in use of drugs. Will be a loner who does not associate with either males or females, and will probably spend a great deal of time in his own home, where he lives alone. Unemployed. Possibly receives some form of disability money. If residing with anyone, it would be with his parents; however, this is unlikely. No prior military record; high school or college dropout. Probably suffering from one or more forms of paranoid psychosis.

As events unfolded, the accuracy of this profile would assume almost eerie proportions, and yet there was a sound basis for each of these observations. Although there had been no obvious sexual assault, experience strongly suggested that the killing was sexual in motivation. This gave the profilers their starting point. At the heart of each profile is a thorough knowledge of criminal statistics. In this case, they knew that data from several years show that most sexual homicides are intraracial—that is, a white killer slays a white victim, or a black killer slays a black victim—and that most sexual killers are white males in their twenties and thirties. Thus the conclusions about race and age were easy to come by. Additionally, the Wallins lived in a white residential neighborhood where any minority figure would have attracted immediate interest.

Illumination about the killer's possible appearance and residence came from the crime scene photographs and police reports. Most killers fall into one of two categories: organized and disorganized. The former are more cunning and are likely to plan their crimes, whereas disorganized killers are creatures of reflex, impulsive and senseless, with little consideration for outcome or consequence. Everything about this murder suggested a random act of violence.

The extent and nature of Theresa's injuries convinced the profilers that her attacker was laboring under a full-blown psychosis and had been for several years. A craving for human blood does not appear overnight; in all likelihood, the symptoms had manifested themselves much earlier, probably warranting some form of treatment. They further reasoned that such a person would care nothing for his own well-being, skip meals, and ignore personal hygiene; hence the remarks about a disheveled and undernourished appearance. A reasonable extension of this hypothesis would presume that the killer lived in squalor, would have extreme difficulty holding down a job, and was quite possibly receiving welfare benefits. As can be seen, the art of psychological profiling is very much like completing a human jigsaw puzzle, except that the compilers have no picture to guide them, only the nuances and likelihoods that experience provides. On one point the BSU investigators were adamant—if not apprehended, the killer was certain to strike again.

Quadruple Murder

Just three days later, their prediction came to pass. Barely a mile from the Wallin household, Evelyn Miroth, thirty-six; her six-year-old son Jason; and Daniel J. Meredith, fifty-two, a family friend, were found shot to death. Also, Evelyn's baby nephew, Michael Ferreira, was missing, presumably abducted by the killer. From the blood-drenched playpen, detectives held out little hope that the infant would be found alive. Of the three bodies at the crime scene, only Evelyn Miroth's had been mutilated. This time the signs of sexual assault were ugly and obvious. Again there were indications that the attacker had drunk his victim's blood. After glutting himself, the killer had made off in Meredith's red Ford station wagon, which was found abandoned a short distance away.

The circumstances of this crime allowed Ressler and Vorpagel to fine-tune their profile. In all probability, the killer had been heavily bloodstained, yet he had abandoned the car in broad daylight and then just walked away. Such recklessness suggested someone even more unhinged than previously suspected, almost certainly a recent psychiatric facility inmate. Statistics inclined them toward the view that the killer probably had a history of fetish burglaries. Of more direct concern to officers pursuing the murderer, however, was the BSU's belief that he lived close to where the car had been dumped.

Using the car as their hub, detectives fanned out in a half-mile radius, asking questions, combing apartment complexes, searching for anyone who might have spotted anything out of the ordinary. In less than forty-eight hours, a young woman reported being accosted by a man whom she knew from high school. She had been so shocked by his emaciated appearance, sunken eyes, bloody sweatshirt, and thick yellow crust around his mouth that at first she had not recognized him. She gave his name as Richard Trenton Chase.

Records showed that Chase lived less than a block from the abandoned station wagon. As the police closed in, Chase attempted to flee. In his hands he held a cardboard box. When tackled by officers, Chase hurled the box at them. Bloodstained papers and rags, a pin from Michael Ferreira's diaper, and pieces of the baby's brain tissue flew out. When finally subdued, Chase was carrying a loaded pistol and Daniel Meredith's wallet.

Chase's apartment was abominable, as predicted. Feces and bloodstained clothing lay strewn about the floor, while in the refrigerator a half-gallon container held body parts and brain tissue. Judging from the contents of three food blenders, Chase had continued to slake his craving for blood. Shortly afterward, the brutalized body of Michael Ferreira was found nearby.

Once in custody, Chase divulged that Theresa Wallin had not been his first murder victim. On December 28, 1977, Ambrose Griffin had just returned from a visit to the local supermarket. As he was carrying groceries indoors, Chase had driven by in his truck and fired two shots. One hit Griffin in the chest and killed him. Analysis of the fatal bullet matched it to Chase's gun.

Bizarre Background

The more investigators delved into Chase's background, the more it highlighted the BSU profile's accuracy. He was twenty-seven and white, and had a long history of sexual problems, fetish burglaries, and drug abuse. Also, his bizarre lifestyle precluded any serious attempt at employment, and he lived alone in his apartment, existing on Social Security benefits.

He became obsessed with the notion that his bodily organs were moving around inside him, turning his blood to powder. His only answer to this perceived assault was to gorge on the fresh blood of others. Such aberrant behavior had led to incarceration in a mental hospital, where he so terrified two staff members that they quit after seeing him chew the heads off birds in the garden. Over the strident protests of those who knew Chase well, a psychiatrist had recommended his release to outpatient status in 1977. Just months later, his killing spree began.

At his trial, Chase remained listless and inattentive. Convicted on six counts of first-degree murder, he was sentenced to death and removed to

San Quentin. Under the taunts of fellow death row inmates, what little sanity Chase had left gave way, and following a report that diagnosed him as "psychotic, insane, and incompetent, and chronically so," he was transferred to California's facility for the criminally insane at Vacaville. There his deterioration continued, until Christmas Eve 1980, when the demons that tormented Richard Chase were finally silenced by a lethal overdose of antidepressant pills, hoarded from his daily medication.

Conclusion

BSU expertise may have forestalled a homicidal catastrophe. Chase had obviously been keeping score. Scrawled on a calendar found in his apartment, on the dates of the Wallin and Miroth-Meredith murders, was the word "TODAY." Forty-four other dates were similarly inscribed throughout the remainder of 1978. Whether Richard Chase intended to kill another forty-four victims is unknowable; that he didn't get the opportunity is largely due to the efficacy of psychological profiling.

John Duffy and David Mulcahy

DATE: 1982

LOCATION: London, England

SIGNIFICANCE: This is the first British murder case in which psychological profiling played a significant role.

In 1982 two men embarked on a four-year series of rapes that cut across London and into the adjoining counties. Eighteen times they struck in unison, although an attack on Barnes Common in November 1984 made it clear that one of the attackers, the shorter of the two, was also carrying out solo rapes. His modus operandi was unmistakable: disarmingly amiable conversation designed to put the woman off her guard, a sudden knife threat, binding the victim's hands with string, then violent rape. Because most of his attacks took place close to railroad tracks, he became known as the Railway Rapist. Victims described him as very short and having "laser eyes" that seemed to bore right through them. The frenzy peaked on one terrifying night in July 1985, when three separate attacks within a few hours convinced police that special measures were necessary. This led to Operation Hart, a manhunt that would become the most comprehensive in Britain since the five-year search for the Yorkshire Ripper that ended in 1981. It was hoped that mistakes made in that investigation—the killer's name cropped up nine different times, but manual cross-referencing had failed to notice it—would be eliminated through computerization, as officers from Scotland Yard joined forces with their colleagues from Surrey, Hertfordshire, and the British Transport Police in a coordinated effort.

The following month, an ex–British Rail carpenter, John Duffy, was arrested and charged with various unconnected violent crimes. Although he was released on bail, his name was entered as a matter of routine in the suspect file of Operation Hart. Meanwhile, the attacks continued.

On December 29, 1985, the Railway Rapist graduated to full-blown murder. Nineteen-year-old secretary Alison Day was dragged from an East London train, taken to a secluded garage in Hackney, and garroted by having a stick inserted through her scarf and twisted. Afterward, her weighted body was dumped in the River Lea. When it was recovered on January 14, 1986, water had obliterated most of the forensic evidence, apart from a few fibers found on her sheepskin coat.

Second Murder

Not until April 17, 1986, did investigators realize that this killer and the rapist were one and the same. Tragically, it took another death to provide the link; fifteen-year-old schoolgirl Maartje Tamboezer was raped and strangled while cycling to West Horsley, Surrey. Diminutive footprints around the body told police that the killer was a small man, but even though he had stuffed burning paper into the vagina, presumably to destroy any forensic evidence, there was still enough semen to show that the killer was blood group A and a secretor. Although he shared this blood group with 42 percent of the population, phosphoglucomutase (PGM) grouping enabled scientists to further isolate the sample so that four out of five suspects would be eliminated. Another clue lay in the distinctive praying attitude of the victims' hands, tied with unusually wide brown string. All of this evidence pointed to the Railway Rapist; from this moment on, Operation Hart became a hunt for a serial killer.

The urgency of the task became terrifyingly evident on the night of May 18, 1986, when Anne Lock, twenty-nine, recently married, and an employee at London Weekend Television, disappeared on her way home from work. Her body was not found until July, but semen traces confirmed that the Railway Killer had struck again.

Now that they knew the killer's blood group, forensic scientists were able to sift through the more than 4,900 men listed in the Operation Hart suspect file, eliminating candidates until they were left with 1,999 names, of whom John Duffy was number 1,505. When interviewed on July 17, 1986, Duffy declined to provide a blood test—as was his right—and immediately had himself committed to a psychiatric hospital for treatment.

At this point, police requested that Professor David Canter of Surrey University, an authority on behavioral science, compile a psychological profile of the killer. Combining statistical analysis of witness and victim statements with his vast insight into the vagaries of human nature, Canter drew up a profile of the Railway Killer that suggested that he lived in the Kilburn area of northwest London, was married and childless, had a history of violence, was plagued by domestic discord, and probably had two close male friends.

When details of the profile were compared with the Operation Hart suspect file, one name leaped from the list—John Duffy. To their fury, the police were now informed that, far from being incarcerated in a mental institution, Duffy was back on the streets. He had signed himself out and had already struck again. This time his victim was a fourteen-year-old girl. At one stage during her ordeal, the girl's blindfold had slipped, affording her a glimpse of the rapist. For some reason Duffy let her live; perhaps he knew his time was up. Whatever the motivation, his Kilburn home became the target of around-the-clock surveillance, leading to his arrest at his mother's house. Seventy items of clothing were removed for analysis by technicians at the Metropolitan Police Laboratory. After more than two thousand experiments, they matched thirteen fibers found on Alison Day's sheepskin coat to one of Duffy's sweaters.

Canter's profile proved accurate in thirteen of its seventeen points. Particularly perceptive was his suggestion that the killer was childless; this had, apparently, been a source of great anguish to Duffy, who had a low sperm count.

At Duffy's trial, his only defense was a feeble claim of amnesia. On February 26, 1988, he was convicted and sentenced to seven terms of life imprisonment.

For years, this case was blighted by one dark cloud—the identity of Duffy's accomplice. This was doubly frustrating for the detectives, as they were convinced that they knew the culprit's identity but were unable to amass the evidence to file charges. The game of cat and mouse with Duffy lasted for a decade. Finally, in a conversation with prison psychologist Jenny Cutler, the Railway Killer let slip a name.

David Mulcahy had been Duffy's friend since childhood. Like Duffy, he'd been born in Ireland, but the two had not met until their respective families moved to north London. The same age, they grew up together and grew vicious together. They were obsessed with martial arts and spent hours watching kung fu movies. They would take potshots at passersby with an air rifle, set fire to woodland, and commit burglaries, all the while urging each other on to more outlandish and brutal crimes. When he was thirteen, Mulcahy pulverized a hedgehog with a plank of wood in the school playground, then trampled on its head. Duffy stood alongside, roaring with laughter.

At the time of the original murder inquiry, Mulcahy, a North London builder, had also figured prominently in police suspicions, but he was made of stronger stuff than his undersized partner and had rebuffed all questions with a sneering disdain. He had watched impassively from the sidelines as Duffy was sentenced to life imprisonment without any hope of release, convinced that the criminal code of silence would be his savior.

But in early 1999 Duffy decided to change the rules. Over a lengthy series of interviews, he painted a terrifying picture of how he and Mulcahy, balaclavas and knives at the ready, would drive around London with Michael Jackson's "Thriller" thundering from the tape deck, hunting down their victims and staking out attack zones with pseudomilitary precision. Once a woman had been snatched, they would flip a coin to decide who would be the first to commit rape. As their sexual cravings became ever more warped, they crossed the line from rape to murder with scarcely a thought. As far as Duffy was concerned, at the time it seemed like a natural progression. Now he just wanted to make a clean breast of it all.

When Mulcahy was arrested in March 1999, this thirty-nine-year-old married man with four children insisted that his former friend was lying through his teeth, prepared to do or say anything to curry favor with the authorities, and he bitterly denied any involvement in either the rapes or the murders. But DNA evidence recovered from the victims' clothing said otherwise. One scientist put the odds of Mulcahy's having been unconnected with two of the rapes at "one billion to one."

When Mulcahy eventually stood trial at the Old Bailey, he listened in malevolent silence as his erstwhile partner spilled the truth about their horrific crime spree. Rarely had murder been so premeditated or so lightly undertaken. On February 2, 2001, Mulcahy was convicted of three murders, seven rapes, and five conspiracies to rape, and ordered to spend the rest of his life behind bars.

Conclusion

Without the psychological profile that led to Duffy, it is likely that many more women would have suffered at the hands of these two sadists. Justice might have been a long time coming in Mulcahy's case, but come it surely did.

IDENTIFICATION OF REMAINS

Whenever a body is discovered, the first question asked by any investigating officer is "Who is it?" Because most murder victims are killed by people they know, the answer to such a question often leads directly to the killer, and so murderers often go to extraordinary lengths to conceal their victim's identity. (The serial killer, slaughtering strangers at random, has little need for such circumspection.) Fortunately, bodies are very difficult to dispose of; bulky and cumbersome, they tend to float in water, resist fire, and smell awful, and they are full of clues as to the person's identity. Of these clues, some such as bones and teeth merit individual consideration and are dealt with elsewhere in these pages, but here we are dealing with the unusual and the inspired. As the reader will find, putting names to a few scraps of remains has frequently provided forensic science with its greatest triumphs.

Martin Thorn and Augusta Nack

DATE: 1897

LOCATION: New York, New York

SIGNIFICANCE: New York's Jigsaw Murder became one of the most infamous crimes in the city's history.

Two boys were swimming in the East River off lower Manhattan on June 26, 1897, when they came across a parcel wrapped in a red oilcloth. Inside was a headless torso. According to the police surgeon, the dissection showed signs of surgical skill. And there was something else: an odd-shaped strip of skin cut from the chest, as though some distinctive feature, such as a tattoo, had been excised. Gradually the river yielded its grim secrets. The lower torso was found several miles upstream in Washington Heights; the final piece in this human jigsaw, the legs, surfaced off the Brooklyn shore. Only the head was never found.

The discoveries caused a sensation, one that George Arnold, crime reporter for the *New York Journal,* was eager to exploit. At the city morgue, he watched the muscular body parts being arranged into something resembling human form. Something about the hands intrigued him; they seemed well cared for yet heavily calloused, a paradox he had noticed previously in masseurs employed at Turkish baths. A survey of such establishments revealed that a masseur named Willie Guldensuppe had not been seen for two days at the Murray Hill Baths. And he had a tattoo on his chest. Arnold traced Guldensuppe to a Ninth Avenue boardinghouse run by a part-time midwife called Augusta Nack. She declared herself baffled by her tenant's disappearance. By this time, the dismembered body had been definitely identified as Guldensuppe by a Dr. J. S. Cosby, who recognized a scar from an operation on one of the dead man's fingers.

Another *Journal* reporter, armed with a photograph of the oilcloth, searched drugstores close to Guldensuppe's home and struck journalistic gold. One store owner remembered selling just such an item to a stout but handsome woman, a description that fitted Augusta Nack. Police persuaded the portly landlady to accompany them to headquarters, where they attempted to browbeat her into admitting all. A detective took her to view the remains, threw back the shirt to reveal the legs, and demanded to know if they belonged to Guldensuppe.

Nack was equal to the challenge. "I would not know," she sniffed, "as I never saw the gentleman naked."

Her response hardly corresponded with what Arnold had been hearing from other residents of the boardinghouse. They reckoned that Augusta Nack and Willie Guldensuppe had been lovers for years. But recently there had been bad blood between them caused by a fellow lodger, a young barber named Martin Thorn.

Confession

At the barbershop where Thorn worked, police heard from another employee that Thorn, an ex–medical student, had confessed to killing Guldensuppe. The way Thorn told it, Nack had seduced him while her regular lover, Guldensuppe, was away, but the bulky masseur had returned unexpectedly and found them in bed together. The beating that Thorn received put him in the hospital and cost him his job at the barbershop. His attempts to disentangle himself from the situation by moving to another boardinghouse failed signally; Nack would not be denied. Neither would Guldensuppe. Again, he caught the couple in flagrante delicto, and again he thrashed the reluctant paramour. For Thorn, this was the final straw. He bought a knife and a pistol, determined to avenge himself.

With news of Thorn's arrest headlining every New York newspaper, police received a call from a Long Island farmer who lived close to where the oilcloth had been purchased. He had a strange story to tell. Apparently, two weeks earlier, he had rented a cabin on his farm to a Mr. and Mrs. Braun. Despite paying fifteen dollars in rent in advance, they had visited the cabin only twice, the second time being on the day that Guldensuppe disappeared. That same afternoon, the farmer noticed that his ducks had turned pink. They had been splashing in a pool of water issuing from the cottage's waste disposal pipe, which had yet to be connected to the main sewer. Closer inspection uncovered what looked to be blood in the water. Now, details of Thorn's arrest had resurrected memories of the mysterious couple.

A police search of the cottage revealed a murder kit: a gun, some rope, carbolic acid, a carving knife, and a saw. This was enough for a warrant to be issued for Nack's arrest. What was already a sensational murder case assumed international significance when Thorn fled to Canada, but he was soon captured and returned for trial.

He was defended by William Howe, half of the notorious New York law firm of Howe and Hummel, two high-priced attorneys whose efforts on behalf of their clients were rarely encumbered by rules of evidence or other tiresome legal considerations. After consideration of all the facts, Howe decided on a full frontal assault: Not only were the two defendants unacquainted with each other, he thundered, but neither of them had met or even known the victim. Drawing on his considerable reserves of rhetoric, Howe went on to ridicule the suggestion that these assorted limbs belonged to Guldensuppe or Golden Soup or whatever the wretched man's name was. They might belong to any corpse. Indeed, what evidence was there that Guildedsoap had ever existed?

If Howe succeeded in planting any seeds of doubt in the jury's mind—and he was without peer in terms of courtroom persuasiveness—then all his good work was entirely undone by Augusta Nack's decision to turn state's evidence. Pleading a fit of remorse (though some suspected that she had been persuaded by the checkbook of William Randolph Hearst, the *Journal*'s owner), she recounted in gory detail every event that took place at the cottage.

Sexual Promises

As suspected, she and Thorn had rented the place with the sole purpose of murdering her violent former lover. Guldensuppe had been lured to the cottage with promises of a renewal in their liaison, but once the brawny masseur stepped inside, Thorn appeared from a closet and shot him in the back of the head. They had dumped Guldensuppe into the bathtub, and while Mrs. Nack went to buy oilcloth, Thorn, the ex–medical student, got to work with his saw. He left the taps running, unaware that the disposal had not been connected to the sewer. Outside, the ducks, rejoicing in this newfound source of water, turned steadily more pink. Nack said that Thorn had encased the missing head in plaster of paris before pitching it into the river.

Such a betrayal forced Thorn into a corner. Hastily he cobbled together his own version of events, one in which he admitted renting the cottage as a "love nest," but he put all of the blame for Guldensuppe's death on his mistress's broad shoulders. He had arrived at the cottage on the day in question to be greeted with the news that Nack had shot her ex-lover. His dismemberment of the body had been solely occasioned by a desire to aid a woman in distress. For some reason, despite this admission, Howe refused to back off from his insistence that this was not the corpse of Willie Guldensuppe, and the bewildering trial drew to its inevitable close. Both defendants were found guilty of murder. Thorn went to Sing Sing, where on August 1, 1898, he became the twenty-seventh person to occupy that institution's least inviting piece of furniture. Augusta Nack plea-bargained her way to a twenty-year jail term.

Conclusion

George Arnold's employer, press baron William Randolph Hearst, was acutely aware of the public's insatiable appetite for homicide and would dispatch armies of reporters on cases such as this one that were likely to boost circulation. His instincts were rarely wrong.

Patrick Higgins

DATE: 1913

LOCATION: Winchburgh, Scotland

SIGNIFICANCE: This identification tour de force was performed by one of the legendary figures in forensic science.

Two Scottish farmworkers, John Thomson and Thomas Duncan, out walking on the afternoon of June 8, 1913, were circling the Hopetoun Quarry, just a few miles west of Edinburgh, when their attention was drawn to a bulky object floating in the water. Their first impression was that someone had tossed a scarecrow from one of the adjacent fields into the water, but a closer look soon disabused them of that notion. The scarecrow turned out to be two small waterlogged bodies lashed together with cord.

Bloated and misshapen, barely recognizable as human, the remains were taken to Linlithgow mortuary for examination by Professor Harvey Littlejohn, head of forensic medicine at Edinburgh University, and his assistant Sydney Smith. This was Smith's first major case in a career destined to span five decades, but rarely, if ever again, would he face such a formidable forensic challenge.

As he cut away the remnants of clothing, his eyes beheld an extraordinary sight—almost all of the body fat had been converted to adipocere, a whitish, suetlike substance caused by prolonged exposure to moisture. Only the feet, encased in boots, were left intact. Adipocere takes between four and five months to develop on the face and neck and a little longer on the trunk if the body is in damp ground; in water its progress is governed by the temperature.

The level of adipocere development that Smith noted here was consistent with immersion in water for at least eighteen months, possibly as much as two years. Only during the autopsy was Smith able to establish that both of the victims were young boys. Measurement of the bones and dental features enabled Smith to say that the taller of the two, at 3 feet 7.5 inches, was between six and seven years old. There was a small injury to the scalp, though it was impossible to state whether this was caused before or after death. The second boy, some five inches shorter, was between three and four years old. Both had close-cropped brown hair, evidently cut shortly before death.

Final Meal

An informative side effect of the development of adipocere was the quite remarkable way in which it had preserved the stomach, so much so that Smith was able to clearly identify what the children had eaten last: green peas, barley, potatoes, and leeks, the ingredients of traditional Scotch broth. Reasoning that the vegetables would have been consumed when fresh—either late summer or fall—and factoring in his earlier estimates regarding the adipocere, Smith estimated that death had most likely occurred in the latter half of 1911. He further thought that the meal had been eaten about one hour before death. This suggested that the boys had lived locally, because the only practical way of getting two victims to such an isolated spot was by foot. While examining the clothing, which was of the cheapest quality, Smith found a faded stamp on one of the shirts. It had formerly belonged to the Dysart poorhouse in Fife. This information enabled the authorities to put names to the corpses.

William Higgins had been born in December 1904, his brother John in 1907. After the death of their mother in 1910, their father, Patrick Higgins, a habitual drunkard and occasional laborer, had applied for their admission to the Dysart poorhouse. Failure to pay for their modest upkeep resulted in his being jailed in June 1911 for two months. On being released, Higgins removed his sons from the poorhouse and dumped them onto a woman named Elizabeth Hynes. Again he neglected his financial obligations, preferring to squander every penny on liquor, so Mrs. Hynes reported him to the Inspector of the Poor. Reminded of his responsibilities and the consequences of failing to meet them, Higgins took his boys away from Mrs. Hynes. At the beginning of November 1911, the two lads vanished.

In conversations with friends, Higgins had made several attempts to explain his sons' disappearance, all contradictory and none plausible.

At his trial in Edinburgh, which began September 11, 1913, his defense was temporary insanity caused by epilepsy, and the main evidence was concerned with proving or disproving this condition. Ultimately the jury adjudged him guilty of murder but with a recommendation to mercy. Nevertheless, on October 2, 1913, Higgins was hanged.

Conclusion

The rarely seen level of adipocere development in this case so impressed Smith that he thought it should be preserved for teaching purposes. He arranged for the specimens to be transferred to the Forensic Science Museum at Edinburgh University, where they have since served as exemplars for generations of budding pathologists.

Hans Schmidt

DATE: 1913

LOCATION: New York, New York

SIGNIFICANCE: This bizarre murder was investigated by one of America's finest early detectives.

On September 5, 1913, a young sister and brother, gazing out from the porch of their Palisades home that overlooked the Hudson River, spotted an unusual bundle being carried along by the early-morning tide. As it bobbed ashore, curiosity got the better of them, and they went to investigate. Inside the manila paper parcel they found a red-and-blue striped pillow. It had been slit open, and among the feathers was the headless trunk of a woman, severed at the waist. The remains were removed to Volk's Morgue in Hoboken, where county physician Dr. George W. King conducted a cursory examination. His first impression was that the dismemberment showed clear signs of a skilled hand. Judging from the softness of the cartilaginous joints, he put the age of the woman at not more than thirty. He estimated her height at approximately five feet four inches and her weight between 120 and 130 pounds, and he judged that she had been in the water a few days at most. (At a later, more thorough autopsy, King found that the woman had given birth prematurely not long before she died.)

The next day, about three miles downriver at Weehawken, New Jersey, two crab hunters came across another parcel, this one containing the lower part of the torso. Wrapped in a newspaper dated August

31, then placed in a pillowcase, it had been weighted down with a large rock. Oddly enough, the composition of this rock was to have a profound effect on the course of the investigation. Geologists determined that it was schist, a grayish-green rock rarely found in New Jersey but very common in Manhattan. It was irregular in shape and appeared to have been broken off by blasting, probably a result of the massive building program then under way in New York.

This revelation raised the question of sovereignty. Bodies found floating in the Hudson River are not entirely uncommon, and usually jurisdiction is decided geographically, that is, on the basis of which side of the river the body washes up on, the New Jersey side or the New York side. But because there was a strong indication that this particular murder had occurred across the river, responsibility for the inquiry passed into the hands of the New York Police Department (NYPD).

Legendary Detective

The investigation was led by one of the great innovators in American criminology, Inspector Joseph A. Faurot. Already he had made his mark. In 1906, following a visit to London, Faurot had returned to New York mightily impressed with Scotland Yard's use of fingerprinting as a forensic tool, and that same year he made history with his arrest of a man with a British accent who was acting suspiciously at a hotel. Despite the man's insistence that he was merely conducting a liaison with one of the hotel's female guests, Faurot sent a copy of his fingerprints to Scotland Yard. Back came confirmation that the man was a notorious hotel thief, thus earning him the distinction of being the first criminal in the United States to be apprehended through the use of fingerprints.

Faurot also understood the value of old-fashioned detective work. In this case, both the pillow and the pillowcase yielded promising clues. On the pillowcase, about an inch high and evidently the handiwork of a novice, the letter *A* had been embroidered in white silk. Such an item, Faurot felt, most likely had come from a lady's boudoir, hinting that the body had been dismembered at a private residence.

A tag on the pillow gave the manufacturer's name, the Robinson-Roders Company of Newark, New Jersey. When questioned, company officials revealed that the pillow had been a disappointing seller; only twelve had been made and all had been acquired by George Sachs, a secondhand furniture dealer. Sachs confirmed that the line had been slow-moving; he still had ten of the pillows in stock. Of the two sold, one was traced to a woman who clearly had no knowledge of the crime; the other had been delivered, together with various pieces of furniture, to an apartment at 68 Bradhurst Avenue. The landlord told Faurot that the flat had been rented two weeks earlier by someone called Hans Schmidt, apparently for a young female relative of his, and Schmidt had ordered the furniture.

All attempts to locate the new tenant foundered, and for five days a team of detectives kept the apartment under surveillance. Still no one showed up. Finally, on September 9, Faurot obtained a passkey and let himself in. One glance was enough to convince him that he had found the murder scene. A dark discoloration on the green wallpaper and another on the floor showed every indication of being blood. Someone had evidently gone to a great deal of trouble to remove them; next to the sink lay a new scrubbing brush and six cakes of soap.

Murder Kit

Inside a trunk, Faurot found a foot-long butcher knife and a large handsaw; both had recently been cleaned. Another trunk held several small handkerchiefs, all embroidered with the letter *A* in the same novice hand as the pillowcase. There was also a bundle of letters addressed to Anna Aumuller. Most were from Germany, but three had return addresses in New York. Faurot interviewed all of the correspondents, a task that ended at St. Boniface's Church on 42nd Street. The priest there remembered Anna Aumuller as a twenty-one-year-old German immigrant who had worked as a servant in the rectory until she was discharged for misconduct. When asked if Anna had known Hans Schmidt, the cleric nodded. He also knew Schmidt's present address—St. Joseph's Church, 405 West 105th Street.

Puzzled more than suspicious, Faurot reached St. Joseph's just before midnight. Father Hans Schmidt answered the door. When Faurot introduced himself, the bulky thirty-two-year-old German-born assistant priest almost fainted. Just minutes later, in a fit of remorse, he unburdened his soul with a bizarre tale of having gone through a form of marriage with Anna—a ceremony he had conducted himself for obvious reasons—only to then kill her, excusing himself on the grounds that "I loved her. Sacrifices should be consummated in blood."

He admitted purchasing the knife and handsaw on August 31, then vacillating until the night of September 2, when he had crept into Anna's bedroom while she lay sleeping and slashed her throat. When questioned about the obvious signs of experience in the dissection, Schmidt acknowledged that he had been a medical student before being ordained. He had discarded the remaining body parts in the Hudson River as well. No more were ever found.

In Schmidt's wardrobe, Faurot found business cards with the name of Dr. Emil Moliere. Schmidt admitted to occasionally posing as a physician under several names, a fact borne out by the discovery of several medicines used to induce abortions. Checking into Schmidt's history, Faurot uncovered a strange mix of religious fervor and con artistry. He had served at several churches across the United States, often arousing suspicion but never facing censure.

At his first trial in December 1913, the jury was deadlocked over Schmidt's sanity. Two months later, a second jury experienced no such misgivings and convicted the priest of murder on February 5, 1914. After a two-year appeal, on February 18, 1916, Schmidt died in the electric chair.

Conclusion

This case was typical of Joseph Faurot's innovative approach to crime solving, an approach that served him well. In time he would found the NYPD's fingerprint department and would later become New York's assistant police commissioner.

Edward Keller

DATE: 1914

LOCATION: Philadelphia, Pennsylvania

SIGNIFICANCE: This crime classic emphasizes the importance of careful observation allied to medical knowledge.

Ten days before Christmas 1915, workers digging in the cellar of a vacant Philadelphia warehouse on Kensington Avenue uncovered a packing case. Inside was a trunk that contained a male skeleton, fully clothed, tall and heavy-boned. Its features, apart from a few scraps of brown and gray hair, had been obliterated with quicklime. Death had resulted from a massive blow to the rear of the head. When the corpse was moved, a dental bridge comprising four teeth fell from its mouth.

The clothing seemed to offer the best chance of identification. A jacket label bore the name of an exclusive Philadelphia tailor, while the pockets surrendered a door key, a prayer book, and a crucifix. When

William Belshaw, a Philadelphia detective who solved several notorious crimes.

asked how long the body had lain buried, the medical examiner hazarded a guess of three years, give or take a few months.

The investigation was headed by Philadelphia's top detective, William J. Belshaw. Right from the outset he had to hack his way through a forest of misleading clues. First, there was the packing case.

Stenciled on one side was a name and address, "John McNamee, Wensley Street," but the 1911 city directory failed to show any such entry. The jacket, too, posed problems. When shown the garment, the tailor whose label appeared inside sniffily dismissed it as mass-produced rubbish, hardly one of his custom-made masterpieces!

If, as seemed likely, the killer had deliberately switched the label to thwart a potential pursuer, then where did this leave the other clues, in particular the dental bridge? Could its integrity be guaranteed? Belshaw had doubts. Sure enough, when he checked the skull, he counted—including the bridge—thirty-five teeth, three more than normal for an adult male. Further discrepancies came to light when the scraps of hair were put under the microscope. The brown hair, fine and silky, was grossly dissimilar to the coarse gray hair. Belshaw reasoned that because the corpse had most of his teeth, this suggested a young man and therefore the brown hair had come from him; the gray hair, he felt, originated from and had been planted by the killer, probably someone middle-aged.

Through local manufacturers of packing cases, Belshaw discovered that on March 14, 1914, a John McNamee had ordered a case for delivery to the 600 block of Wensley Street. But this was only twenty-one months prior to the discovery in the basement, a much shorter length of time than the three years suggested by the police surgeon. Belshaw blamed the abundance of quicklime for the doctor's miscalculation as to how long the body had lain in the pit.

The address on Wensley Street turned out to be a rooming house whose landlady recalled one of her tenants receiving such a case. She thought his name was McNamee and described him as mid-twenties, tall, with blond hair and prominent teeth. Belshaw produced the key found on the corpse and began trying every door in the building. On the second floor he found a lock that it fitted.

Records for 1914 showed that the warehouse where the body had been found had previously been occupied by the Red Star Laundry, now defunct. The ex-proprietor, Edwin Klempner, confirmed to Belshaw that McNamee had worked for him as a night watchman until March 1914, when he had abruptly vanished. Proceeding on the basis that the likeliest motive for murder had been robbery, Belshaw surveyed every local pawnshop. One such establishment reported that in March 1914 they had loaned ten dollars on a watch to a Mr. John McNamee of Wensley Street. The ticket had never been redeemed and they still had the watch. Inscribed on its case was a name—Daniel McNichol.

Fearful Wife

A 1914 directory listed McNichol's address as 866 North 22nd Street. When Belshaw called, a timid-looking woman admitted that she was McNichol's wife and that he had vanished on March 30, 1914, but she refused to enlarge on why she had not reported his disappearance. Her description of him—tall, well-built, brown hair, good teeth—fitted the corpse perfectly, as did her statement that McNichol, a devout Catholic, never went anywhere without a prayer book and crucifix. As gently as he could, Belshaw informed Marie McNichol that her husband had been found.

Through a flood of tears, Mrs. McNichol explained her previous reluctance to report Daniel's disappearance. Apparently, one of his business partners, a man named Edward Keller, had called with news that Daniel had absconded with fifteen hundred dollars in company funds. Keller said that if Mrs. McNichol was prepared to make good the deficiency, he would not contact the police. Dutifully the poor woman repaid every cent that her husband had misappropriated. She had no idea where Keller lived—he had always called for his money—but described him as about forty-five, with deep-set eyes.

Belshaw soon found those eyes staring out from a police file photograph; they had an oddly familiar look. After a moment's thought, the mist cleared—Keller and Edwin Klempner, boss of the Red Star Laundry, were one and the same! Keller's criminal record went back almost a quarter of a century and included a fourteen-year jail term for embezzlement. Known accomplices included a twenty-six-year-old nephew, Albert Young, who in every detail fitted the description of John McNamee. Belshaw speculated that the two men had lured McNichol to the laundry and killed him. But that still left one loose end—the dental bridge. Where had it come from? The likeliest solution was provided by Young's mother in New York, who said that her son wore just such a bridge. Belshaw pondered the likelihood of Young's voluntarily relinquishing his bridge and could reach only one conclusion.

Edward Keller readily confessed his identity when Belshaw came to call; he also admitted that he and Young had killed McNichol because they owed him money. But on one point Keller would not budge: he emphatically refuted any suggestion that he had killed Young as well. His erstwhile partner, he said, had cut and run.

When it came time to stand trial, Keller repudiated this confession, claiming coercion and declaring himself the victim of a plot hatched by his missing nephew. Enough doubt was created for the jury to return a verdict of manslaughter. Keller received a twelve-year jail sentence. Albert Young was never seen again.

Conclusion

This tale has a curious conclusion. During the eight years Keller spent behind bars, he appeared to reform, and on his release, he married a social worker, who found him a job as night watchman at the Corn Exchange National Bank. About one year later, on the morning of December 20, 1925, an elderly man, panting hard and clutching a satchel, hailed a cab in midtown Philadelphia and asked to be driven to the railroad station. He kept urging the driver to go faster. At the station, the cab drew to a halt. The driver, puzzled by the silence behind him, turned around. Edward Keller was sprawled across the backseat, dead from a heart attack. Beside him lay the satchel, stuffed with more than six thousand dollars in bills stolen from the bank just hours before.

Abraham Becker and Reuben Norkin

DATE: 1922

LOCATION: New York, New York

SIGNIFICANCE: Not only are stomach contents useful for establishing time of death, but here they provided indisputable proof of identification as well.

For years, Abraham Becker and his wife Jennie had endured a bitter, loveless union, plagued with violent quarrels. Finally, in 1920, Becker left his wife and four children and eloped with his longtime mistress, twenty-four-year-old Anna Elias. But after three months in Cleveland, they returned to the Bronx, and Becker moved back in with his wife. This heralded such a dramatic improvement in the Beckers' domestic situation that on the night of April 6, 1922, no one was surprised to see the couple at a friend's party with Abe playing the attentive spouse, plying Jennie with canapés, grapes, figs, and almonds.

On the way home, their car developed engine trouble and chugged to a halt. Becker, a chauffeur and experienced mechanic, raised the hood to diagnose the problem, then called for Jennie to help him. As she climbed from the car, he felled her with a wrench, raining blow after blow on her head. Certain she was dead, Becker hauled her body to a nearby grave he had already prepared, an ash pit. For added insurance, he doused the body with lime. Then he drove home.

Questions from inquisitive neighbors were parried with the story that Jennie had run off to Philadelphia; another man, he said, had spirited her away. Few believed this, however; Jennie was a devoted mother and would never willingly abandon her brood. Becker, on the other hand, incited a torrent of local gossip by depositing the four children in local orphanages, thus clearing the way for Anna Elias and their two-year-old daughter to move in with him. If his actions were rash, then his tongue was positively suicidal. Often he could be overheard boasting to friends, "Congratulate me. I have got rid of my wife!"

Such comments inevitably led to his downfall. In November, the garrulous chauffeur was arrested as a material witness in the disappearance of his wife. Even in jail, his tongue refused to stay still, as he bragged to a visiting acquaintance that he had killed his wife and "buried her so deep they [police] couldn't find the body in a hundred years." Had Becker realized that the person listening so avidly to these admissions was an undercover police agent, he might have been more circumspect.

Murder Plot

During these conversations, Becker made frequent mention of his business associate, Reuben Norkin, an auto shop owner. Norkin was quickly arrested and grilled. At first, he denied the allegations, but then he admitted helping Becker bury his wife, although he insisted that his complicity stopped at just that. Norkin led detectives to where Jennie Becker's body lay in the yard of his auto shop. After digging down four feet, the search party found a rotting corpse.

When confronted with the remains at the morgue, Becker was dismissive. "My wife's a bigger woman than that."

"What about the clothes?" asked the police.

"These aren't the clothes my wife was wearing." Becker's insistence that his wife had been wearing low-heeled shoes, unlike the fashionable high heels worn by this corpse, prompted retrieval of the original missing-person report he had filed. In it he had described exactly the clothes worn by this dead woman. Still, he refused to budge from his assertion that these were not the remains of his wife.

To settle the argument once and for all, the corpse was turned over to medical examiner Dr. Karl S. Kennard for autopsy. The crushed skull obviously indicated severe blows to the head, but from the condition of the bronchial tubes, it appeared as though Jennie Becker had actually been alive when buried and had died from suffocation. Also, the lime had not totally destroyed the stomach, which meant that when Dr. Alexander O. Gettler analyzed the contents, he was able to find traces of grapes, figs, almonds, and meat-spread sandwiches—the very items Becker had lovingly fed his wife at the party. Becker, obviously shaken by these discoveries, blustered that any woman could have eaten such food, until Gettler administered the coup de grâce: examination of the meat spread found it to be identical to the canapés served by the party hostess—prepared according to an old family recipe.

Becker thought fast. He now admitted that the corpse was that of his wife but claimed that Norkin had killed her to "get even with me, because we had a row over an automobile." When asked why he had not reported the murder, Becker replied, "because he [Norkin] has killed a lot of other people and would kill me, too." Becker then added the rather astonishing assertion that after killing Jennie, Norkin had considered his account settled, and the two men had become firm friends again!

Not surprisingly, Norkin's version of events was an exact mirror image of Becker's, even down to fearing that Becker would kill him, too. Norkin said that Becker had first broached the subject of killing his wife months earlier. "In April he asked me to lend him a shovel and I let him have one. . . . Later on he told me that he had buried his wife in a pit near my shop."

While Becker's guilt was beyond dispute, prosecutors opted to seek the maximum penalty for Norkin as well. In separate trials, both men were found guilty of first-degree murder, and each went to the electric chair in 1924.

Conclusion

Dr. Alexander O. Gettler was for many years head of New York's Chemical and Toxicological Laboratory in the medical examiner's office. His expertise and lucid delivery made him an excellent witness in dozens of trials, but rarely did he exceed the heights reached in this memorable case.

Patrick Mahon

DATE: 1924

LOCATION: Eastbourne, England

SIGNIFICANCE: Sir Bernard Spilsbury often cited this case as the most challenging he ever encountered.

For some time, Jessie Mahon had been suspicious of her husband's frequent absences from home. Knowing better than most the effect his salesman's smile and glib tongue had on the opposite sex, she feared he might be up to his old philandering ways again. Finally, propriety gave way to curiosity, and she began emptying the pockets of his many suits. Her attention was drawn to a baggage-check ticket from Waterloo train station in London. She asked a friend, an ex–railroad policeman, to investigate. On May 1, 1924, he presented the ticket at Waterloo and was given a Gladstone bag, the contents of which sent him running to Scotland Yard. They shared his concern but thought that justice could be best served by replacing the ticket in Mahon's pocket and allowing him to reclaim the bag.

The next day, when Mahon breezed into the baggage-check office, a waiting police officer stepped forward. Mahon cheerfully agreed that the bag's blood-soaked contents—a cook's knife and a monogrammed case, both of which had been sprinkled with disinfectant—did look ominous, but attributed the stains to some dog meat he had carried recently. Told to try again, because the blood was human, Mahon crumpled. In a whisper, he admitted that the initials on the case—EBK—were those of his mistress.

At thirty-seven, Emily Kaye was in love, engaged to be married, intending to emigrate, and pregnant. The only cloud on her horizon was that the father of her child, Patrick Mahon, already had a wife. There is no evidence to suggest that Emily was greatly perturbed by Mahon's matrimonial state; on the contrary, she seems to have been quite ruthless, making plans for their imminent departure to South Africa. Unfortunately, she accepted Mahon's invitation to spend the weekend of April 12 at a bungalow he had rented on a desolate stretch of Sussex coastline known as the Crumbles.

Mahon had prepared for Emily's visit by purchasing a large cook's knife and a tenon saw, which showed quite remarkable prescience, because during the course of an argument that weekend, Emily fell and accidentally struck her head on the coal scuttle with fatal results, leaving Mahon to marvel that he had just the right equipment to ensure that his embarrassment did not become public knowledge. Emily, he said, or what was left of her, was at the bungalow in Sussex.

House of Horror

When Scotland Yard detectives, accompanied by Sir Bernard Spilsbury, entered the deserted house, they saw nothing that was immediately recognizable as Emily Kaye. What they did find was detailed by Spilsbury in his later report: two large saucepans of boiled human flesh, saucers and other receptacles swimming with greasy human fat, thirty-seven portions of flesh in a hatbox, a fiber trunk holding large chunks of torso including a heavily bruised shoulder, and a biscuit tin full of various organs, all atop a blood-drenched carpet.

Spilsbury spent eight hours in this hellhole, sifting through the fireplaces and dustpans until he had recovered a thousand fragments of calcined bone, much of it almost dust. Every item was cataloged and removed to his laboratory in London for reassembly. The task took him several days, usually after his colleagues had gone home, to spare them the harrowing spectacle of Emily Kaye being reassembled. He found everything except the skull and a portion of one leg. All had been part of a female body, pregnant at the time of death.

Although unable to state categorically how Emily Kaye died, Spilsbury had seen enough to demolish Mahon's claim of an accident. The idea of the flimsy coal scuttle inflicting such massive shoulder bruising and lethal injuries to the skull in a single fall, without either buckling or at least being dented, was preposterous. And neither had she been carved up with the blunt knife that Mahon claimed he had used, a frantic and futile attempt to deflect attention away from the deadly purchases that he originally claimed had been bought long afterward. Spilsbury always suspected that Mahon had bludgeoned Emily Kaye to death with a missing ax handle, which like the skull, was never found.

Allegedly, while on remand, Mahon told another inmate of burning the head on a stove in the middle of a thunderstorm; the intense heat had caused the eyes to open, which had so terrified Mahon that he had rushed from the house. Coincidentally, during his cross-examination on this very topic, a sudden clap of thunder rumbled overhead. Mahon cringed in terror, arm raised defensively, yellowish face staring skyward. Nothing else needed to be said. He was executed on September 9, 1924.

Conclusion

An interesting outcome of this case arose through Spilsbury's abhorrence at seeing police officers at the bungalow move scraps of rotten flesh with bare hands. Practically no specialized equipment—rubber gloves, tapes, or fingerprint powder—then existed for use at a crime scene. To address this deficiency, Scotland Yard, in consultation with Spilsbury and other forensic experts, assembled the famous Murder Bag that became an essential feature of every such investigation.

Henry Colin Campbell

DATE: 1929

LOCATION: Cranford, New Jersey

SIGNIFICANCE: Inspired detection led first to the identity of a murder victim, and then to her heartless killer.

Studious, soft-spoken, and apparently a man of some substance, Richard Campbell seemed to be everything that Mildred Mowry, a middle-aged widow, was looking for in a husband. On his application to the matrimonial agency through which they met, the sixty-one-year-old New Yorker had listed his occupation as doctor; she believed him, and after a whirlwind courtship, the couple married on August 28, 1928. The next day, Campbell suggested that she deposit her life savings of one thousand dollars in his bank account so that she might receive the benefit of his deft investment skills. Mildred gladly agreed.

What should have been honeymoon bliss turned out to be less-than-romantic solitude because of Campbell's sudden announcement that his surgical work required his immediate presence in California. Leaving Mildred to languish in the small Pennsylvania town of Greenville, he traveled not to the West Coast, but to Elizabeth, New Jersey, and the bosom of his genuine wife and family, whose existence he had naturally kept from Mildred. He still kept in touch with her by letter, though, blaming his continual absences on too much work and not enough money.

Mildred, upset by Campbell's furtive behavior, eventually traced the elusive doctor to an accommodation address on West 42nd Street, New York. The proprietor, Louis Mirel, explained that Campbell had not used the facility for some time and suggested another address. After no luck there, either, Mildred returned home. On February 22,

1929, out of the blue, Campbell suddenly turned up in Greenville and informed Mildred that he was taking her to New Jersey with him. They set out in his dark-painted roadster. It was after midnight when Campbell halted just outside the town of Cranford. As Mildred lay dozing on the seat beside him, Campbell took careful aim with a .38 automatic and fired one bullet through the crown of her head. Afterward, he dragged the body from the car, doused it with gasoline, and set it alight, intending to destroy all evidence and possibility of identification.

At 5:30 the next morning, a truck driver spotted the still-blazing corpse and rushed to fetch the police. They rolled the charred body in the snow to extinguish the flames, but it was burned almost beyond recognition. To aid in identification, Public Prosecutor Abe J. David engaged Pinkerton's Detective Agency. William F. Wagner was assigned to the case, and he faced a thankless task. After tearing out the labels, the killer had bundled the victim's clothes around her head and set fire to them. What was left of them, poor in quality and old, suggested someone of modest circumstances. A few trinkets of jewelry still adorned the body, eliminating robbery as a motive. The shoes were manufactured by a St. Louis company, which reported that large quantities of the cheap brown oxfords had been shipped to the coal regions of Pennsylvania.

Death Mask

Determined that he would not be defeated, Wagner had the undertaker construct a death mask of the victim. The completed likeness, with its unusually high cheekbones, inclined Wagner to think that the victim had been of Polish extraction. Because there were numerous Polish settlements in western Pennsylvania, he had flyers circulated in that region, giving the victim's approximate age and describing her clothing. It wasn't long before the Greenville chief of police wired Wagner with a promising lead.

Just recently, two women, Mrs. S. D. Staub and Mrs. H. G. Dodds, had contacted him, worried about the disappearance of their friend Mildred Mowry. Six months earlier, she had told them of her marriage to a surgeon, Dr. Richard Campbell, but now the two ladies were agitated because they had not heard from her for two months. When they described Mildred as wearing special arch supports in her size 6C shoes, Wagner confirmed that these half-burned brown oxfords were similarly equipped.

In early April, copies of Mildred Mowry's medical records were given to the New Jersey authorities. Like the murder victim, Mowry had one finger stiffened at the joint, and this, together with charts from her dentist, convinced Wagner that he had identified the victim. In Greenville, a search of Mildred's apartment revealed letters from her absent husband, some giving the Mirel accommodation address. But the agency, although recalling the uncommunicative Dr. Campbell, knew nothing of his present whereabouts.

Wagner went to Elkton, Maryland, where the bigamous marriage had taken place. County records gave the doctor's address as 3707 Yosemite Street, Baltimore, which turned out to be an empty lot. However, the tenants of 3505 just down the street disclosed that their landlord's name was Henry Colin Campbell of Westfield, New Jersey, only a few miles from the murder scene. Police files showed Campbell, a convicted forger, with a history of marital irregularity, littered with wives and annulments. The Westfield house was deserted, but Wagner learned that Campbell's mail was being forwarded to 471 Madison Street, Elizabeth, New Jersey.

Drug Addict

On April 11, detectives called at the Elizabeth address and were met by Campbell's unsuspecting and legitimate wife. She told them that her husband was not due back until nine o'clock that evening. On the stroke of nine, a gray-haired man ambled under the nearby streetlight and turned in to the house. Wagner and another detective moved forward to make the arrest. In Campbell's pocket, they found an automatic, later identified as the murder weapon. Campbell, who described himself to police as an advertising man, meekly allowed himself to be taken away. Within an hour, he had signed a confession to the murder of Mildred Mowry, although he later claimed that he could remember nothing of the killing, saying he had been in a drug-induced stupor at the time. What percentage of Campbell's actions were attributable to morphine addiction is unknown, but several physicians later testified that his health was in decline and that he required ever-increasing doses of medication.

In all probability, Mildred Mowry was not Campbell's first victim. Just one year earlier, middle-aged governess Margaret Brown suddenly left her job to marry a doctor she had met through a matrimonial agency, taking with her seven thousand dollars in savings. Her body, shot and burned in identical circumstances, was found just fifteen miles away from where Mowry met her death. While the similarities were marked, it proved impossible to fix the blame for that murder on Campbell. In the end, it was all academic. On June 13, 1929, Campbell was found guilty of first-degree murder. He was executed the following April.

Conclusion

In times of such limited communications, the sheer size of the United States was often the criminal's greatest ally. A crime committed in one state, unless of national significance, stood very little chance of being reported elsewhere. Without William Wagner's brilliance and determination, it is entirely conceivable that Campbell would have evaded justice, and given his track record, he very likely would have gone on to kill again.

Patrick Brady

DATE: 1935

LOCATION: Sydney, Australia

SIGNIFICANCE: What became known as Australia's Shark Arm Murder has achieved legendary status in the annals of forensic science.

In mid-April 1935, two fishermen caught a fourteen-foot tiger shark off the coast of Sydney, Australia. Reluctant to kill such a magnificent creature, they gave it to the Coogee Beach aquarium, but the shark never settled into captivity, refusing to eat and lolling sluggishly in the pool. On April 25, the shark suddenly went berserk, thrashing about in a frenzy that culminated with it vomiting up a huge mass of waste, among which was a well-preserved human arm! The arm was Caucasian in origin, muscular and brawny, and had an unusual tattoo depicting two boxers squaring up to each other. Knotted tightly around the wrist was a length of rope.

Aquarium experts were amazed that the shark's digestive juices had not dissolved the limb. Normally, such flesh would have been destroyed within thirty-six hours, but apparently the trauma of capture had so affected the fish that its digestive system had shut down. This suggested that it must have swallowed the arm very shortly before being captured a week earlier. After the shark was killed, an autopsy of the unfortunate creature failed to find any more human remains.

The skin on the fingers was still intact, but it was in very fragile condition. Scientists at the Sydney police laboratory managed to remove it carefully in small flakes and reassemble it, eventually obtaining a set of faint but definite prints. A check of police records had already revealed an alarming incidence of tattooed people reported missing, but the investigation soon homed in on James Smith, a forty-year-old ex-boxer with a criminal record. His wife identified the arm as that of her husband; the identification was later confirmed by fingerprints.

Smith had vanished two weeks before, on April 8, nine days before the shark was captured. He had told his wife of renting a seaside cottage for a fishing holiday with Patrick Brady, a denizen of the Sydney underworld with strong links to forgery and drug-trafficking circles. Taken into custody, Brady denied all knowledge of Smith's disappearance, but he did suggest that a wealthy boatbuilder named Reginald Holmes might be able to throw some light on the mystery.

Convinced that they were dealing with a case of gangland murder, detectives strove resolutely to locate the rest of the victim, as legal minds were already debating whether the discovery of just one arm constituted definitive proof of death. Even the armed forces joined in, as navy divers explored the sea bottom and air force planes made low sorties along the beaches. But they found nothing.

By chance, Professor Sydney Smith of Edinburgh University in Scotland, one of the world's foremost pathologists, was visiting Australia at this time and was asked to assist the local forensic medicine team in examining the arm. All agreed that it had been severed at the shoulder joint with a sharp knife, not bitten off by a shark. The amputation also seemed to have been carried out some considerable time after death, thus indicating that the arm's owner had not committed suicide.

Murder House

They next examined the seaside cottage. According to the owner's inventory, a small tin trunk, a mattress, three mats, and a length of rope were missing. This paved the way for a tentative reconstruction of the crime in which James Smith was killed, hacked apart on the mats, then squeezed into the tin trunk. The mutilated parts that the trunk would not accommodate, including the tattooed arm, were then roped to the outside, and the whole lot was dumped far out at sea. Somehow, the arm had worked loose, then to be swallowed by the unfortunate shark, which had suffered such gastric discomfort in the aquarium.

The rest of the story was equally bizarre. Three days after Brady's arrest, Reginald Holmes—who had denied even knowing Brady—was captured by a police launch in Sydney Harbor after a high-speed chase. His face was covered in blood from a slight bullet wound, the legacy, police thought, of a failed suicide attempt. Not so, said Holmes, insistent that someone had tried to kill him. He also now claimed that Brady had killed James Smith and had disposed of the body.

Holmes was right about someone wanting to kill him. On the very eve of the inquest into Smith's death at which he was to be a key witness, someone shot him twice in the head. His body was found in his car, late at night, under the Sydney Harbor Bridge. The roar of the traffic passing overhead had probably drowned out the sound of the gunshots.

Brady, of course, was in custody and therefore had a perfect alibi. Testimony at the inquest dragged on for twelve days until Brady's lawyers obtained a Supreme Court injunction to stop the proceedings on the grounds that one arm did not prove that James Smith was dead. Furious, the police responded by charging Brady with murder. It was a rash gesture. In court, Brady coolly accused yet another man, Albert Stannard, of involvement in Smith's death. The jury, no doubt influenced by the Supreme Court's earlier decision, decided that there was insufficient evidence to warrant a charge of murder, and Brady was set free. Later, Stannard and John Strong, both suspected hit men for the Sydney underworld, were tried for shooting Holmes, but after two trials they, too, were acquitted.

From what is known of the main characters, it appears likely that all this mayhem and mystery was the result of a struggle for control of drug trafficking in that part of Australia.

Conclusion

Discussing this case much later in his career, Sydney Smith wrote, "It is a trite saying but true that fact is stranger than fiction."

Buck Ruxton

DATE: 1935

LOCATION: Lancaster, England

SIGNIFICANCE: This forensic watershed was replete with radical new techniques that determined both corpse identification and time of death.

A young woman out walking on the morning of September 29, 1935, was crossing a bridge near Moffat, Scotland, when she spotted numerous parcels, some wrapped in newspaper, littering the bank of the River Annan below. One, she realized to her horror, was a human arm. A more thorough search of the stream and its banks yielded seventy pieces of human body parts. Included in this total were two disfigured heads. The remains were taken to Edinburgh for analysis.

The pathologists charged with piecing together this grisly jigsaw puzzle—John Glaister, professor of forensic medicine at Glasgow University; Sydney Smith, New Zealand–born professor of forensic medicine at Edinburgh University; and J. C. Brash, professor of anatomy at Edinburgh University—were united in one opinion: Whoever carried out the mutilation did so with a high degree of anatomical skill and considerable patience (they estimated that the dissection had taken eight hours to complete). All of the customary identifying marks—sex organs, fingertips, and facial features—either had been obliterated or were missing. Despite the profusion of parts, the team believed that they were probably dealing with two bodies at most. One head was definitely female. The other skull appeared to be that of a male, but the extreme mutilation made identification almost impossible. This confusion led the first newspaper reports to list the victims as a woman of twenty-one, about five feet one inch tall, and a man of about sixty, five feet six inches in height.

Even without medical evidence to help them, Dumfriesshire police were able to establish roughly when the remains had been dumped in the gully. It could not have been before September 15, because that was the date of one newspaper used to wrap some of the remains. Conversely, the discovery of more body parts several hundred yards downstream suggested a date no later than September 19, when the stream had last been in full spate following two days of heavy rain.

Acting on this assumption, police narrowed their inquiries to anyone reported missing in the few days prior to September 19.

By chance, the chief constable of Dumfriesshire had been reading an account in a Glasgow newspaper of the disappearance from Lancaster, one hundred miles to the south, of a Mary Jane Rogerson, a maid at the home of Dr. Buck Ruxton. A telephone call by the chief constable to his Lancashire counterpart brought forth the information that Ruxton's wife had also disappeared at about the same time. Putting two and two together, the Scottish policeman asked the pathologists to reexamine the supposedly male bones to determine whether they were those of a strongly built female. He also asked Lancaster police for a description of Mrs. Ruxton.

Stormy Relationship

Ruxton, an Indian-born doctor, and Isabella Van Ess, although never legally married, lived together in Lancaster as man and wife and produced three children. Their union was forever teetering on the brink of collapse, blighted by Ruxton's insane, often violent jealousy that frequently led to interventions by the police. And now there was good reason to believe that that point of collapse had been reached. Ever since the disappearance of Isabella and Mary Jane Rogerson—who were last seen on September 14, 1935—Ruxton had been careering around Lancaster like a madman, seemingly determined to put a noose around his own neck. He had fabricated all manner of stories to explain the women's disappearance, none of them convincing, and had hounded the local police with demands that they search his home so that he might quash the vicious rumors circulating that he had done away with both wife and maid. At first the police had been inclined to dismiss Ruxton as a crank; after the request from Scotland they weren't so sure.

North of the border, other clues began to surface. Foremost were the newspapers used to wrap the remains. One was a torn copy of the *Sunday Graphic,* dated September 15. An officer, examining a fragment of the paper, found a mutilated headline that read "—ambe's Carnival Queen–rowned." An accompanying photo showed a young girl wearing a crown. Contacting the paper, he learned that this was a special regional edition, sold only in the vicinity of Morecambe and Lancaster.

Once a local newspaper vendor confirmed that a regional edition of the *Sunday Graphic* had indeed been delivered to the Ruxton household on September 15, the circumstantial evidence began to steamroller.

Rompers used to wrap one of the body parts were identified by a woman who had given them to Mary Rogerson's mother for onward transmission to the Ruxton children.

Finally, the police took Ruxton up on his suggestion that they search his home at 2 Dalton Square. Despite evidence of a lengthy and exhaustive cleanup, they still found bloodstains everywhere, and the drains contained unmistakable traces of human fat. Ruxton's bluff had failed. Summoned to the police station on October 12, he was questioned through the night, and the next morning, at 7:20 A.M., he was formally charged with murdering Mary Rogerson.

Two Bodies

In Scotland, the pathologists continued their labors, as they assembled the skulls, torsos, seventeen parts of limbs, and forty-three portions of soft tissue into something resembling human form. In shape and size, the first body conformed exactly with Mary Rogerson. The bones and teeth indicated a woman approximately twenty years old, Mary's age exactly. The second body was that of a woman estimated to be between thirty-five and forty-five. Isabella was thirty-four years of age.

Curiously, it was more what the pathologists didn't find that convinced them that they were on the right track. Certain distinguishing features on both bodies had been eliminated. Mary was known to have a squint; the eyes of the first body had been removed from their sockets. Also, the skin on the upper arm had been scraped away, right in the area where Mary had a prominent birthmark. Similar disfigurement covered the spot where Mary's appendix scar would have shown. Likewise, the area at the base of her thumb, where an old injury had left a scar, had been shaved of tissue. Most convincing of all, fingerprints from the first body matched those found at Dalton Square on articles habitually handled by Mary Rogerson. It was evidence good enough to stand up in any court. Just about the only thing that the pathologists couldn't ascertain was the actual cause of death; the mutilation was too thorough.

The second body was equally illuminating. Several of the teeth had been wrenched out, and the nose was missing completely. Isabella Ruxton had very prominent teeth and a large nose. Also, her legs from the knees to the ankles were conspicuously of the same thickness; on this body the legs had been pared of all flesh. In this case, though, the condition of the lungs and tongue left no doubt that asphyxiation had been the cause of death. A fractured hyoid confirmed that she had been throttled. Interestingly, the eyes, nose, ears, lips, and tips of the fingers—all other areas in which evidence of asphyxia are found—had been removed, adding to suspicion that the murderer had specialized medical knowledge.

However, conclusive proof that this was Isabella Ruxton came as the result of a major forensic breakthrough. For the first time, the photograph of a victim was superimposed over the skull to see if it would match. It fitted perfectly. Between them, the trio of Glaister, Smith, and Brash had slotted the final piece into an astonishing forensic jigsaw.

In order to establish when the two women had been murdered, yet another forensic first was employed. Dr. Alexander Mearns of Glasgow University, in a pioneering feat of medical detection, by studying the life cycle of the maggots that infested the remains on the riverbank, was able to establish that the victims were killed at about the time that Isabella Ruxton and Mary Jane Rogerson were last seen alive.

On November 5, 1935, a second count of murder was formally levied against Ruxton, that of killing his wife. He was convicted on both counts and sentenced to death. Following his execution on May 12, 1936, a newspaper released a full confession made by Ruxton in which he admitted killing Isabella in a jealous fury, only to be interrupted by the unfortunate Mary Jane. As a witness, she, too, had had to die.

Conclusion

The ramifications of this dual murder were so far-reaching that Brash and Glaister felt compelled to record their findings in a book, *Medico-Legal Aspects of the Ruxton Case* (1937). Lucid and scrupulously detailed, like the subject matter with which it deals, this book, too, has achieved the status of a classic.

Arthur Eggers

DATE: 1946

LOCATION: San Bernardino, California

SIGNIFICANCE: This case is an example of how postwar California put the pursuit of criminals on a sounder scientific basis.

In January 1946, a visitor to the San Bernardino mountains was exploring one of that region's desolate box canyons when he spotted something lying beneath a broken tree branch. A closer look revealed it to be a middle-aged woman's corpse, half-naked, with the head and hands hacked off. She had been shot twice with a .32 caliber pistol. Preliminary inquiries by the local sheriff's department were hampered by the mutilation; without a face or fingerprints, identification was next to impossible, and neither the corpse's clothing nor a bloodstained tartan blanket found nearby offered any clues. Even so, fortune was with the investigators; only luck had led to the gruesome discovery, and the victim had been dead for less than ten hours. In different circumstances, the corpse might not have been found for months, by which time weathering and wild animals would have reduced it to a pile of unrecognizable bones.

Despite extensive newspaper coverage of the murder, no clues as to the woman's identity were forthcoming, which left only the missing-person files as a possible source of the corpse's identity. A three-week search produced the name of Dorothy Eggers. She had been reported missing by her husband, Arthur, at about the time the body was found, and at forty-one, she certainly sounded like a possibility, but the file described someone much slimmer and taller than the victim.

When filing the report, Arthur Eggers, a fifty-two-year-old clerk, had emphasized his wife's riotous lifestyle. He recounted how she would leave him and their two adopted children and hitch rides to nearby towns in search of other men, returning only when the money or the man had run out. There seemed to be genuine sadness in his eyes when he suggested that perhaps she had found someone willing to take her permanently.

Eggers's story certainly sounded tragic, but it did nothing to explain why Eggers had given such a blatantly inaccurate description of his missing wife, for already detectives knew that in height and stature, Dorothy Eggers and the body found in the mountains were identical. Eggers, summoned to the mortuary, removed all doubt by identifying the headless corpse as that of his wife, and corroboration came from the family doctor who had treated Dorothy for a large swelling on the foot. The bunion was still evident.

Bloodstained Car

Although there was considerable suspicion against Eggers, not a shred of evidence existed to connect him with the corpse. And then fate lent a hand. Recently, Eggers had sold his car. The buyer, a deputy sheriff, had noticed some brownish spots in the trunk that looked like dried blood. Aware of their possible significance, he reported his findings to his superiors, and Eggers was arrested. After days of questioning, he finally admitted that he had murdered his wife during an argument over her promiscuity. He had grabbed his pistol and fired, and then had disposed of the body while his children were out at the movies. About the head and hands he would say nothing other than that he vaguely remembered burning them.

As anticipated, Eggers soon recanted this confession, saying that it had been obtained under duress, and he fell back on his initial claim that Dorothy had probably been murdered by one of her casual lovers. This left the San Bernardino prosecutors in a quandary. They needed more than a retracted confession to win in court; hard physical evidence was necessary to link Eggers to the crime. To find that evidence, they turned to the Los Angeles Police Department, some sixty miles to the west. On the West Coast, they were unrivaled in the scientific approach to detection.

Criminalist Ray Pinker was assigned to the case, and the wealth of trace evidence he uncovered was astonishing. His first efforts, though, met with disappointment; fragments of bone found in the Eggers incinerator turned out to be of animal origin. More encouraging were the strands of hair Pinker found on the corpse, which matched samples from Dorothy's hairbrush. Next, he examined those items found with the body. Eggers's children had already identified the tartan blanket as coming from their home; now Pinker matched blood flecks on the blanket to the stains the deputy sheriff found in the trunk of Eggers's car. Both were of the same type as the victim. Examination of the blanket using ultraviolet light failed to reveal any semen stains, thus reducing the likelihood that Dorothy had first engaged in sex and had then been murdered by a casual lover.

Pinker was asked to analyze a .32 caliber pistol found at the Eggers family home. He test-fired the gun into a wooden container packed with cotton waste, then retrieved the bullets. Under a comparison microscope, they were indistinguishable from the murder slugs, demonstrating that this was the gun that had killed Dorothy Eggers. Even more deadly was a handsaw that Eggers had used around the house. Embedded in its teeth, Pinker found minute particles of human blood, flesh, and bone.

With such an overwhelming forensic case, Eggers's trial in the summer of 1948 became something of a formality. Convicted of murder, he died in the gas chamber at San Quentin on October 15, 1948.

Conclusion

Ray Pinker's outstanding contribution to the solution of this murder made headlines all across the state and provided many Californians with their first glimpse into a world that had previously been the stuff of fiction.

Richard Crafts

DATE: 1986

LOCATION: Newtown, Connecticut

SIGNIFICANCE: Trying to convince a jury that murder has taken place when no body has been found has never been easy. But the astonishing wealth of forensic evidence gathered in this case should have been enough to convince the most hardened skeptic.

Darkness had already fallen when Helle Crafts, a Pan Am flight attendant, and her co-worker, Rita Buonanno, left New York and drove north. They had just returned on a flight from Germany and were hurrying home, eager to beat the expected early winter snowstorm. At around 7 P.M., Rita dropped Helle off at her house in Newtown, Connecticut. Helle Crafts waved good-bye and hurried indoors. No one outside the house ever saw her alive again.

That night, November 18, 1986, five inches of snow fell, bringing down power lines and tree branches. All the next day, Rita tried to phone Helle but got no answer until the evening, when Richard Crafts picked up the phone. The forty-nine-year-old Eastern Airlines pilot, icily indifferent, disclaimed all knowledge of his wife's whereabouts. Rita was alarmed. She knew from talking to Helle that Richard Crafts was unpredictable—dangerously so—and she knew that Helle had recently initiated divorce proceedings on account of his adultery. But most of all, she remembered that Helle had once said, "If anything happens to me, don't think it was an accident."

Rita's apprehension proved contagious. Other friends pestered Crafts nonstop on the phone, until he eventually snarled that Helle had flown to Denmark to visit her sick mother. However, when contacted, Helle's mother said that she was in fine health and had not seen her daughter for some time.

On November 25, Marie Thomas, nanny to the three Crafts children, spoke of a dark, grapefruit-sized stain she had seen three days earlier on the master bedroom rug. Kerosene, Crafts had said, but now that rug was missing, along with two others. (It was later learned that on November 22, Crafts had ordered carpeting worth fifteen hundred dollars from a local store.)

News of Helle Crafts's disappearance was greeted coolly by the local police. Their circumspection was understandable; domestic dislocations are an everyday occurrence, and most missing persons turn up alive and well. Even so, on this occasion many wondered whether official slowness to investigate had anything to do with Richard Crafts's work as an auxiliary police constable. Certainly Helle's friends thought so, and they kept up their campaign.

Finally, on December 2, Detective Harry Noroian contacted Crafts, who explained that Helle had left the house early on November 19. An hour later, he had gone to his sister's house in Westport. Around lunchtime he returned, but Helle was still not back, and he had not seen her since. He presumed that she had run off with an "Oriental boyfriend" from Westchester County, New York. When Noroian suggested that a polygraph might stave off the rising tide of rumor, Crafts jumped at the chance and passed with flying colors.

The first indication that all might not be right with Richard Crafts's version of events came from a credit card receipt. It showed that on the afternoon of November 19, he had purchased bedding. Yet when talking to Noroian, he had emphatically denied leaving the house after returning from Westport. This prompted a closer examination of Crafts's credit card records, and a particularly odd sequence of purchases, which had begun almost a week before Helle's disappearance.

On November 13, he had put one hundred dollars down on a Westinghouse chest freezer, telling the store clerk that he would collect it later. He returned for the freezer on November 17. The following day at 1 P.M., just hours before Helle's return from Europe, Crafts rented another item, one he had reserved four days earlier—a wood chipper.

Gruesome Theory

Nothing had prepared the investigators for this. The implications of the purchase of a wood chipper were almost too horrible to countenance. A wood chipper's blades are designed to shred large branches; their effect on a human body was unthinkable. Grim confirmation of this nightmarish scenario came when a witness told of seeing a truck and wood chipper parked on a bridge over the Housatonic River just after dark on November 20. Detectives now had to face the ghoulish but very real prospect that Richard Crafts had first killed his wife and then disgorged her remains into the river.

On December 26, detectives carrying a search warrant entered Crafts's house. Because Crafts was not present, and to refute possible later claims of improper search or evidence planting, the entire exercise was videotaped. The first thing that struck detectives was the size of Crafts's gun collection: fifty weapons in all, a small armory. But the rest of their cursory inspection proved disappointing, and it was not until the arrival of Dr. Henry Lee, head of Connecticut's forensic laboratory, that significant progress was made.

In the master bedroom, Lee applied his considerable skills to a brown smear on the mattress. When sprayed with ortholotolidine solution (tolidine, ethanol, glacial acidic acid, and distilled water), the stain turned blue, indicating that it was blood. Next came the problem of deciding how it had come to be there. Dr. Lee, an internationally regarded authority on blood spatter analysis, determined that the blood had landed directly on the mattress, with medium velocity, at an angle of about ten degrees, suggesting that if Helle Crafts had been murdered in this room, then she had not been lying on the bed. Beyond this, Lee would not yet commit himself. Later laboratory tests confirmed that the blood was human, that the stain was relatively recent, and that the blood was of the same type as Helle Crafts's—O positive.

When yet another witness came forward to tell of seeing a wood chipper on the banks of the Housatonic in the early-morning hours of November 19, detectives concentrated their attentions on a section of the river known as Lake Zoar. There, in a culvert, they found an envelope addressed to Helle L. Crafts. It was the first clear evidence linking her to the area.

The Search

Slowly the net began to close in around Richard Crafts. While Lee supervised the laboratory analysis, a team of divers scoured the Housatonic. From deep within its icy waters they recovered a Stihl chain saw; nearby was the serrated cutting bar. These were sent to Elaine Pagliaro, an assistant of Lee's. Embedded in the bar's teeth she found human tissue, hair, and blue fibers that matched fibers at Crafts's house. Although the chain saw's serial number had been filed off, scientists were able to restore it and match the saw to one that Crafts had bought.

The evidence gathering was not confined just to the river. In a car that Crafts had driven, Lee found among loose wood chips in the trunk traces of human flesh, hair, and bone fragments.

On January 10, 1987, a weeklong search of the snow-covered riverbank began. It was an enormous undertaking. Searchers had to thaw the snow inch by inch and then sift through the soil on hands and knees. In all, the search yielded fifty-nine slivers of human bone, part of a finger, five droplets of blood, two tooth caps, 2,660 strands of hair, three ounces of human tissue, and two fingernails. It wasn't much—Lee estimated it at roughly one thousandth of a human body—but it provided enough material for more than fifty thousand forensic tests. Every single hair was examined microscopically and found to be human and blond, like Helle Crafts's. The presence of the roots indicated that they had been chopped from a skull.

Albert Harper, University of Connecticut anthropologist, deduced from the fat content of the bones that they were just months old. When he ground some fragments to powder to allow the production of antibodies and antigens, they revealed the blood type O positive—the same as Helle Crafts's.

But the most conclusive evidence came from one of the tooth caps. Comparison with dental records positively identified it as belonging to Helle Crafts.

In the midst of all this technical wizardry, Crafts was arrested and charged with his wife's murder. Prosecutors theorized that he had killed Helle to prevent a financially ruinous divorce. After clubbing her to death in the bedroom, they speculated, he stored her body in the freezer to facilitate easier carving with the chain saw. Then came disposal of the human parcels using the wood chipper. It was a grisly scenario but quite consistent with the facts.

On July 15, 1988, Crafts's first trial ended in a deadlock when one juror held out for acquittal and a mistrial was declared. The following year, on November 21, 1989, Crafts was finally convicted of murder and sentenced to fifty years of imprisonment.

Conclusion

Working with barely a handful of human matter, scientists were able to prove beyond any reasonable doubt that these were the mortal remains of Helle Crafts. As her husband stood spraying those remains into the Housatonic River, it could never have crossed his mind just how heavily those few scraps would ultimately weigh on the scales of justice.

SEROLOGY

Bodies are leaky objects; once punctured they tend to ooze or spray blood indiscriminately. This is hardly surprising, because the average human being has about ten pints of blood gurgling through his or her system at any given time. Although it was recognized as early as 1875 that there were various types of blood, it was not until 1901 that Karl Landsteiner, an Austrian-born biologist, standardized the grouping system and gave it its modern form. By separating the serum from red blood cells in a centrifuge, then adding red blood cells from different people to the serum, he found that two distinctively different reactions occurred. In some cases, the serum seemed to attract the red blood cells, but in others it was repelled. One group of cells agglutinated—or clumped together—and the other didn't.

Landsteiner labeled these two blood types as A and B, but he soon realized that there was a third type that didn't react the same way as either A or B but showed characteristics of both. This he called C, although it soon became known as O. One year later, an assistant discovered yet another type of serum that did not agglutinate with either A or B. This one was called AB. Thus the four major blood groups were identified.

In the mid-1920s, Landsteiner discovered another grouping system as a result of injecting rabbits with human blood. These groups he identified as M, N, and MN (a combination of M and N). In 1930, he was awarded the Nobel Prize for Physiology or Medicine. Landsteiner went on to discover the Rhesus factor in 1940 as a result of experiments with rhesus monkeys.

The next great advance in understanding blood types came in 1949, when two British scientists observed that the nuclei in cells of female tissue usually contain a distinctive structure that is rare in males. It was called the Barr body, after one of its discoverers, and it is most noticeable in white blood cells. The presence of this female-only structure is accounted for by the differences in chromosomes between males and females.

These advances have had a profound effect on criminology. When blood is found at a crime scene, investigators can now determine whether it is indeed blood by performing the Kastle-Meyer test, which uses a solution of phenolphthalein that turns pink when it comes into contact with even the minutest trace of blood. Then, by using the precipitin test (see the Ludwig Tessnow case on page 241), investigators establish whether it is animal or human blood. Finally, they determine the blood group and the sex of the person who spilled that blood. For all these reasons, blood is one of the most reliable and informative types of evidence, and, as we will see, its impact on crime solving has been immense.

Pierre Voirbo

DATE: 1868

LOCATION: Paris, France

SIGNIFICANCE: This case established the reputation of one of France's greatest detectives.

Following complaints from customers about the quality of his water, the owner of Lampons, a Parisian restaurant in the rue Princesse, investigated the basement well. Floating on the surface was a malodorous parcel wrapped in cloth. Gingerly, the owner fished it out and was horrified to discover that it contained the lower portion of a human leg. On January 26, 1869, he reported his discovery to the Sûreté, and a new recruit named Gustave Macé was sent to investigate. Macé found yet another parcel floating just below the surface—a second leg, encased in part of a stocking.

Both limbs had been sewn into pieces of black calico, about a yard square, tied at each end. Macé thought he detected the hand of a seamstress or tailor in the bags' skilled construction. Also, someone had monogrammed the letter *B* in red cotton into the stocking. All the medical examiner could say was that the legs appeared to be female and that they appeared to have been in the water for a fairly long time—perhaps six weeks. Macé began checking through the file of missing women, 122 in all. He had painstakingly reduced the number to 84 when he received unwelcome news: A second doctor, more experienced in forensic medicine, had examined the legs and declared them to be male. All of Macé's efforts thus far had been in vain. Undeterred, he merely redirected his efforts toward finding a missing male.

He remembered that in mid-December a male human thigh wrapped in a blue sweater had been recovered from the River Seine. At around the same time, another thighbone had turned up in the rue Jacob. Two days later, a laundry proprietor reported seeing a stocky, mustachioed fellow scatter pieces of meat from a basket into the Seine. Then, just before Christmas, two gendarmes had stopped a similar-looking man in the early morning hours not far from Lampons. He was carrying a large hamper. Inside were several parcels. He opened one and showed them a ham. Satisfied, the gendarmes had let him go. Macé suspected that this confrontation had so unnerved the killer that he had dumped the legs in the first available place, the well at Lampons. It also suggested that he lived close by.

The concierge of the restaurant building knew of no tailor currently living there but directed Macé to a seamstress who once had. Mathilde Dard was now a nightclub singer; she was also the ex-mistress of a tailor and political firebrand named Pierre Voirbo, who happened to match the description of the nocturnal meat distributor. Mathilde had lost touch with Voirbo but suggested that Macé try a Madame Bodasse, the aunt of one of Voirbo's friends. Macé, alert to the possibility that the letter *B* might refer to Bodasse, took the elderly woman the monogrammed stocking. She instantly identified her own handiwork; also, she recognized the sweater as belonging to her nephew, Desiré Bodasse, who lived in the rue Mazarine.

Murder Site?

Macé found Bodasse's apartment deserted, but a flickering candle and the fully wound clocks suggested that someone was still living there. Madame Bodasse also confirmed that a batch of negotiable securities was missing from a secret drawer. Macé ordered a watch put on the apartment. For more than a week nothing happened. Then, out of the blue, Voirbo breezed in to see Macé at the Sûreté. He oozed confidence, due in large part to his background as a police informer, and he listened to what Macé had to say with amused indifference. Some of his composure slipped, though, when Macé produced a rent receipt from Voirbo's former apartment—he had paid the bill with one of the missing bonds.

Macé placed Voirbo under arrest and then searched his current lodgings from top to bottom. There were stacks of newspapers, all containing accounts of the discovery of the legs in the restaurant, and in the cellar he found two barrels of wine. Inside one, concealed in a cylinder, were the missing securities.

A feature of Gallic criminal investigations is the predilection for recreating crime scenes. Macé already had a fair idea of where and how the crime had taken place; now he wanted confirmation. Collecting Voirbo and several officers from the station, the convoy journeyed to the house in the rue Mazarine, where Macé had noticed that the tiled floor in one room had a prodigious slope. With a theatrical flourish, Macé poured a carafe of water onto the tiles, then watched it gather in a pool beneath the bed. Voirbo remained tight-lipped as Macé ordered that the tiles be removed one by one. As each tile came up, dried bloodstains could clearly be seen on the sides and underneath.

Realizing that the game was up, Voirbo broke down and confessed. He had battered Bodasse with a flatiron and then had slit his throat. Afterward, dressed only in his underwear, he had dissected the body. He thought that he had cleaned up the mess thoroughly but had failed to notice that recess beneath the bed. His motive had been simple greed; he wanted the securities because he would soon be getting married.

Ever resourceful, Voirbo managed to make one brief dash for freedom but was soon recaptured. While he was in jail awaiting trial, friends brought him gifts. One, a loaf of bread, contained a razor. He used it to cut his own throat.

Conclusion

In highlighting Gustave Macé's contribution to this case, it should be remembered that if not for the intervention of the second doctor, Amboise Tardieu, this murder would probably have never been solved.

Ludwig Tessnow

DATE: 1901

LOCATION: Rugen, Germany

SIGNIFICANCE: The killer had previously escaped justice, but a landmark advance in forensic science prevented a repeat performance.

Off the northern coast of Germany, in the Baltic Sea, lies the island of Rugen. Here, on July 1, 1901, two young brothers from the village of Gohren, Hermann and Peter Stubbe, eight and six respectively, left their home to go out to play. When they did not return, the family feared the worst. The following morning, a search party found dismembered body parts scattered over a wide area of local woodland. Tracking the bloody trail led them to the disemboweled remains of the Stubbe brothers.

A suspect was soon in custody. Ludwig Tessnow, a jobbing carpenter from the neighboring village of Baabe, had been seen talking to the children on the day they disappeared, but he denied any involvement in their deaths. A routine search of his home turned up boots and clothes all bearing dark stains and all recently washed. Tessnow's explanation sounded reasonable enough, considering his job: The stains were wood dye, he said, used daily by countless craftsmen in his line of business.

Tessnow was taken before the examining magistrate at Greifswald, Johann Schmidt. For some reason, Schmidt doubted the carpenter's tale but didn't know why. Then the mist cleared. Three years earlier, German newspapers had reported a startlingly similar case in the town of Osnabruck, several hundred miles to the west. Two young girls, seven-year-old Hannelore Heidemann and her friend Else Langemeier, eight, had been found butchered in a wood near their home on September 9, 1898. A man seen loitering near the woods on the day of the murder had been detained because his clothing was heavily stained. He gave his name as Ludwig Tessnow. But when he told the police that the stains were simply wood dye from a recent job, he was eliminated from their inquiries.

Magistrate Schmidt, together with the local prosecutor, Ernst Hubschmann, continued digging. They discovered that just weeks before the brutal Stubbe slaying, on June 11, a farmer had seen a man fleeing from one of his meadows. Puzzled, he went to investigate. Seven of his sheep had been hacked to pieces, their parts strewn around the field. When Hubschmann arranged for the farmer to view a lineup of possible suspects, he had no hesitation in picking out Tessnow as the culprit.

Revolutionary Test

Tessnow maintained his innocence; the stains on his clothing, he insisted, were wood dye, nothing else—which left Hubschmann in a quandary. Apart from these stains, there was nothing to connect the carpenter with any crime. But just as he was on the verge of freeing Tessnow, he learned that a young German biologist, Professor Paul Uhlenhuth, had recently developed a revolutionary new technique that could differentiate not only between blood and other stains but also between human blood and the blood of other animals.

Uhlenhuth had found that after he injected protein from a chicken's egg into rabbits and then mixed the rabbit serum with egg white, the egg proteins separated from the clear liquid to form a cloudy substance or precipitate. An extension of this process led to the production of rabbit-based serums that would precipitate, and identify, the proteins of the blood of any animal, including humans.

Since that time, the method has been greatly simplified. A liquid sample of suspected human blood is placed on a glass slide treated with gelatin next to a similar sample of the biological reagent. When an electric current is passed through the glass by means of electrodes, the protein molecules in the two samples filter outward through the gelatin toward each other. If a precipitin line forms where the antigens and the antibodies meet, this indicates that the sample is human blood.

After almost a week spent examining a pair of overalls taken from Tessnow's house, together with other items of clothing, Uhlenhuth submitted his report on August 8, 1901. It turned out to be a momentous day in the annals of forensic science. The overalls were indeed stained with wood dye, but he also found seventeen traces of human blood. Tessnow's suit and shirt yielded similar results, although his jacket was found to be stained with sheep's blood. After his trial and a lengthy confinement, Tessnow was executed at Greifswald Prison in 1904.

Conclusion

The precipitin test is extremely sensitive, requiring only minute blood samples. Positive results have been obtained on human blood that has been dried for as long as fifteen years, and experiments on tissue samples of mummies several millennia old have proved successful.

Jesse Watkins

DATE: 1927

LOCATION: San Francisco, California

SIGNIFICANCE: This historic case involved the use of ultraviolet light to detect bloodstains.

A thick fog enveloped San Francisco as Jesse Watkins headed toward that city's historic military post, the Presidio, on the night of August 21, 1927. Watkins, twenty-four, an ex–stable hand, had recently been fired for laziness by the Presidio stable master, Henry Chambers. Now he was going to get the $340 of pension money that was owed to him. Watkins broke in and shook the old man awake, only for Chambers to grab a revolver and begin firing. One bullet out of the three struck Watkins, lodging in his cheek. With blood streaming from the wound, he wrested the revolver away and then used it to club Chambers repeatedly. When the old man was still, Watkins bound and gagged him with his own underwear, only to realize that Chambers was dead. He then set about ransacking the apartment, searching for cash. After pocketing a few dollars, he burrowed his head down through the fog, grateful for its cover because his clothes were heavily bloodstained.

Upon reaching his Lombard Street apartment, he told his roommate, Sergeant Harry Edwards, that he had been mugged by a gun-toting

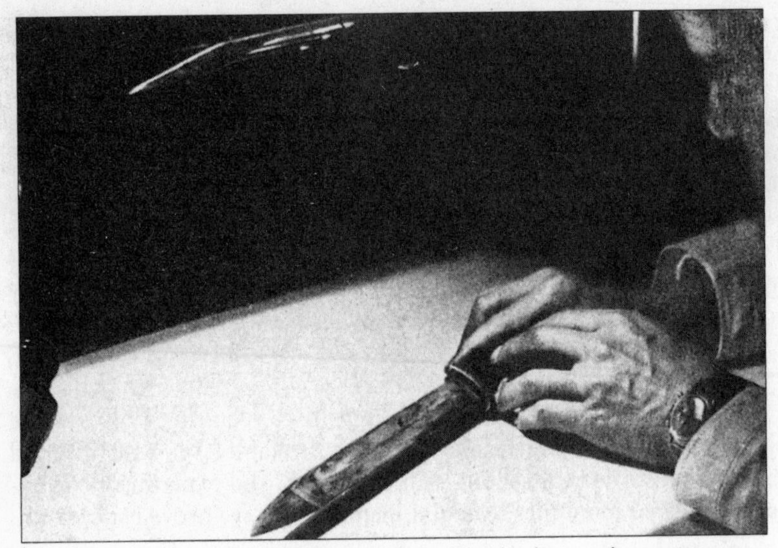

Here, ultraviolet light is used to detect bloodstains on a bayonet.

thief. Edwards was skeptical but removed the bullet with tweezers. Watkins, aware that news of the bloody killing would soon be all over San Francisco, scrubbed his shirt to get rid of the evidence and sent it to the laundry to make sure.

He was right; the murder dominated local headlines, and all the San Francisco police had to go on was a heel imprint left in the dust on Chambers's floor. Within the print were four letters—*VERE*. Routine questioning led them to another Presidio employee, who mentioned Jesse Watkins as a man with a grudge against Chambers. When detectives went looking for Watkins, they found Edwards. His story about the bullet was enough for them to place Watkins under arrest. A search of the suspect's belongings revealed a pair of old shoes, which were forwarded to Edward O. Heinrich at the University of California for examination.

Vital Clue

First, he noted that because of the wearer's manner of walking, both rubber heels were more worn on the inner side than the outer. Even so, the letters *VERE*—identical to those found in the heel imprint on the floor of Chambers's room—were clearly visible. Microscopic examination of those heavily worn parts revealed the faint outlines of two more letters—*RE,* which had preceded the others until being almost obliterated by the shoe owner's odd gait. This gave Heinrich the full word, *REVERE,* and the trade name of the heels. Although any reasonable person would have to conclude that Watkins had left the print in the dust, there was nothing to say when that print had been made, as any competent defense counsel would be quick to point out.

Heinrich wasn't finished yet. By chance, Edwards had kept the pellet extracted from his roommate's cheek. Using the recently invented comparison microscope, Heinrich was able to demonstrate conclusively that the slug had been fired from Chambers's gun. Finally, he turned his talents to the laundered shirt. In a darkened room, he spread the shirt out carefully on a table and then placed ultraviolet-light equipment in position. The sleeves showed nothing, which didn't surprise Heinrich, as many people roll them up. But all over the front, greenish-blue spots began to appear, unmistakable signs of blood.

The case came to trial in October 1927. Heinrich's evidence demolished defense alibi claims and their attacks on so-called circumstantial evidence. Found guilty of murder and sentenced to life imprisonment, Watkins later confessed to the killing but declared that it had been accidental; he had intended only to beat Chambers up, not to kill him. His final comments on the matter included a bitter reference to the shirt and the damage it had done him: "If I'd known about those violet lights, or whatever you call them, you can be sure I'd have burned the blamed thing up."

Conclusion

Had this murder occurred just a few years earlier, Watkins would probably have walked free. Instead, he found, as thousands have since, that old-fashioned crime is no rival to the latest laboratory advances.

Jeannie Donald

DATE: 1934

LOCATION: Aberdeen, Scotland

SIGNIFICANCE: In this case, microbiology came to the assistance of serology and helped capture an unsuspected killer.

Eight-year-old Helen Priestly lived with her parents in a grim three-floor tenement building in Aberdeen. At half past one on the afternoon of April 20, 1934, her mother sent her to buy bread from a nearby store. When she had not reappeared after some time, Mrs. Priestly went out looking for her. Staff at the bread shop confirmed that Helen had bought a loaf just after half past one. She was last seen walking homeward at 1:45 P.M.

All of Mrs. Priestly's tenement neighbors volunteered to help in the search except Alexander and Jeannie Donald, a morose couple who kept themselves apart from other tenants. The search, which lasted all night, ended at five the next morning when Helen's body was found near the ground floor communal lavatory. And yet half an hour earlier, when another neighbor had used that same lavatory, the body had not been there.

Dr. Robert Richards, the police surgeon, found Helen lying in a sack, fully clothed except for her underwear. Rigor mortis had set in, and there was vomit on her dress, possibly a result of the cinders that had been thrown into the sack, some of which had lodged in her mouth. Although the body was on its right side, postmortem lividity showed that it had lain on its left for several hours after death.

Blood on her tights and dress and grievous injuries to her genitals indicated that she had been savagely raped. Because of this, and because the body had obviously been dumped by someone who had easy access to the lavatory, police questioned every male in the building. Meanwhile, Dr. Richards and Theodore Shennan, professor of pathology at Aberdeen University, began the autopsy.

Death had resulted from asphyxia. Vomit was found in the windpipe and the smaller air tubes, while bruises typical of manual strangulation mottled the voice box and windpipe. Digestive changes to Helen's last meal of meat and potatoes, eaten at half past twelve, indicated that death had occurred not later than two o'clock, a time confirmed by a roofer working nearby who had heard a girl scream at 2 P.M. What the pathologists found next—or rather didn't find—turned the investigation on its head. Despite the deep penetration and tearing of the sexual organs, no semen was present; they adjudged that all of the injuries had been inflicted before death by a sharp implement, by someone attempting to simulate rape.

Female Killer?

Proceeding on the assumption that the killer lived in the tenement, investigators were stymied by the fact that every male resident had a cast-iron alibi—unless the killer was a female who had attempted to mask her involvement by faking the sexual assault. Easily the strongest suspect was thirty-eight-year-old Jeannie Donald, who lived directly below the Priestlys. She admitted feuding with the dead girl's family but said that at 2 P.M. she was out shopping. However, when it was learned that a shop she claimed to have visited had been closed that day—and after the police found what appeared to be blood spots in her apartment—Jeannie Donald, along with her husband, was arrested. As it happened, the stain proved not to be blood, but the police still felt they had enough cause to hold Jeannie Donald, although her husband's alibi proved unshakable and he was released. At this point Professor Sydney Smith of Edinburgh University was asked to assist the inquiry.

He began with the sack. It was of Canadian make, used to ship cereals to Aberdeen, and had a hole in one corner. Five similar sacks were found in the Donalds' apartment, and all had holes in one corner where they had hung from a hook. Inside the murder sack Smith found the cinders and a small amount of household fluff containing human and animal hairs. The human hair had not come from Helen Priestly; it was a different color, much coarser, and artificially waved. When compared with hairs from the head of Jeannie Donald, Smith found them to be indistinguishable.

However, this was far from conclusive evidence, so Smith turned his attention to the rest of the detritus. It contained traces of cat and rabbit hair, along with fibers of wool, cotton, silk, linen, and jute dyed different colors—more than two hundred fibers in all. When compared with similar samples from the Donald household, using microchemical and spectroscopic analysis, twenty-five different fibers were matched, including the human and animal hair. Examples taken from other apartments in the building failed to produce a single match.

But the strongest evidence was yet to come. A closer scrutiny of the Donalds' apartment uncovered bloodstains on two washcloths, a scrubbing brush, a packet of soap flakes, a piece of linoleum, and two newspapers dated the day before the crime. All were group O, the same as the victim's. Smith circumvented defense objections to Jeannie Donald's giving blood by obtaining one of her used sanitary towels from prison. This showed a different blood group entirely, so obviously the stain had not come from her. Unfortunately, the victim's blood group was shared by almost half the population, gravely limiting its forensic usefulness.

Bacterial Contamination

Building on this serological bedrock, Smith reexamined the victim's wounds. Suddenly a thought occurred to him: Might not the rupturing of Helen Priestly's intestinal canal have led to her blood being contaminated by bacteria? Gathering up the victim's clothing, together with the bloodstained articles from Jeannie Donald's household, Smith consulted his colleague Thomas Mackie, professor of bacteriology at Edinburgh University. Mackie found considerable evidence of bacterial contamination, including one highly unusual strain that he had never encountered before—except on a washcloth from the Donald apartment.

The trial of Jeannie Donald began on July 16, 1934. Much of the testimony was given over to medical and scientific evidence. While none of the laboratory findings were conclusive in themselves, the sum total was devastating. Smith produced 253 separate and diverse items for the prosecution, a colossal performance that even the defense counsel had to grudgingly praise. The jury took only eighteen minutes to return a guilty verdict.

Donald's death sentence was later commuted to life imprisonment, partly, it is believed, because of Smith's belief that she had not deliberately intended to kill the child. It was known that Helen Priestly was in the habit of baiting Mrs. Donald with the name "Coconut," a reference to her frizzy hair. Smith theorized that Helen had insulted Jeannie Donald as she passed her door and that the woman had caught hold of Helen and shook her. Abruptly the girl had lapsed into unconsciousness. (At the postmortem, it was discovered that Helen suffered from an enlarged thymus gland—a condition likely to cause rapid and unexpected collapse.) Thinking that the girl had died, Jeannie Donald panicked and pulled her indoors, where she decided to simulate rape with the sharp instrument. The excruciating pain awoke Helen, who screamed and vomited, some of which she swallowed. This may have been the cause of death, though the bruises suggested that Donald had strangled her to be sure. Then she had stuffed the body into a sack, and awaited her opportunity to dispose of the body in the dead of night.

Conclusion

Between the wars, Professor (later Sir) Sydney Smith was an untiring pioneer of the multifaceted approach to crime scene analysis. Unlike other, more renowned pathologists of the time, he avoided the traps of intransigence or dogma, preferring to allow the facts to speak for themselves, always willing to try the latest techniques. With his death in 1969, forensic science lost one of its finest and fairest practitioners.

Joseph Williams

DATE: 1939

LOCATION: Bournemouth, England

SIGNIFICANCE: Confronted by skilled advocacy, a jury repudiated the scientific evidence and set a man free. But was it the correct decision?

It was approaching midnight on the night of May 21, 1939, when the body of Walter Dinivan, a sixty-four-year-old widower, was found in the living room of his apartment in Bournemouth on England's south coast. His skull had been crushed. Later that night he died in the local hospital without regaining consciousness. The autopsy, performed by Sir Bernard Spilsbury, indicated that the killer had first attempted to strangle Dinivan and, when that had failed, had finished him off with a torrent of hammer blows to the head.

Chief Inspector Leonard Burt of Scotland Yard studied the crime scene thoroughly. Everything smacked of robbery as the apparent motive: A small living room safe had been emptied, as had Dinivan's pockets; also, his rings, watch, and gold chain were gone. On the floor lay a brown paper bag, crumpled and twisted, which Burt suspected had been wrapped around the murder weapon. The room yielded a rich crop of varying fingerprints. Comparison with relatives eliminated all of the prints except one—a thumbprint lifted from a toppled beer glass.

One of the odder discoveries was a hair curler found on the floor. Flushed with embarrassment, Dinivan's grandchildren suggested it might be related to the old man's fondness for entertaining prostitutes. But they had no explanation for the cigarette butts strewn across the sofa and carpet. Burt, aware of recent advances in saliva examination, ordered all of the butts gathered up for analysis. In the meantime, he interviewed local prostitutes. Several knew Dinivan as a regular client, but all dismissed the idea of using such an old-fashioned hair curler. During these conversations the name of Joseph Williams, a septuagenarian crony of the dead man, cropped up. Normally strapped for cash, around May 21 he had suddenly come into some money.

This was confirmed by a background check: Since the murder, Williams had paid off considerable arrears on his mortgage. When Burt called on the old man, he was met by a vile-tempered brute, toothless and unkempt, with thick glasses. After attributing his recent affluence to a win on the horses, Williams launched into a fearful diatribe against Dinivan, who had recently refused his request for a small loan. Included in the bilious onslaught were several tactless references to the dead man's safe. When asked if he minded providing an example of his fingerprints, Williams angrily ordered Burt from his house.

Blood Secretor

The inspector's return to the station was greeted by good news—the smoker of the cigarette butts was a secretor. In 1925, it had been discovered that some 80 percent of the population secrete their specific blood group information in other bodily fluids, such as saliva. This enabled Home Office analyst Roche Lynch to identify the cigarette smoker's blood group as AB, the rarest type, found in only 3 percent of the population. What was needed now was a specimen from Williams; given his attitude, this would be no easy task.

Burt's solution was both simple and ingenious. Officers kept Williams under permanent surveillance, with orders to contact Burt immediately should the subject ever enter a pub. A few days later the call came through. Burt rushed around to the pub, ostensibly on a social visit. Williams greedily accepted Burt's offer of a drink and a cigarette, and over the next hour or so, his attention diverted by alcohol and talk of horse racing, contrived to fill an entire ashtray with cigarette butts. When he finally weaved his uncertain way to the door, Burt watched him go, then gathered up the ashtray's contents. The next morning they were dispatched to Lynch.

His report confirmed that Williams was indeed a secretor, blood group AB. Burt decided it was time to turn the screws and confront the old man. Williams angrily insisted that Burt search his house, declaring he had nothing to hide. It was a monumental blunder. In a coal shed was a bundle of brown paper bags, the kind that had been found in Dinivan's living room. Williams was scornful. "I used to be a greengrocer," he sneered. "All greengrocers have bags."

At this point Williams, his confidence inflated to reckless levels, boldly proffered his hands, daring Burt to take his fingerprints. Burt gladly accommodated him and was later rewarded with the news that Williams's right thumbprint matched the print found on the glass. It was enough to arrest the vitriolic old man.

Despite a compelling circumstantial case against Williams—his sudden affluence, an admitted animosity toward Dinivan, the thumbprint, and the saliva test—there was still no direct evidence to link him with Dinivan on the night of the murder. Even so, men had been hanged on considerably less, and Williams's attorney, Norman King, decided that his client's only chance of avoiding the noose lay in an all-out assault on the credibility of the saliva test. (Just months earlier another jury had rejected saliva evidence and acquitted a defendant.) Dismissing all of the other evidence as irrelevant, King held aloft one of the cigarette butts and asked the jury, how was it possible to determine a blood group from invisible saliva traces? Could they, in all conscience, send a man to the gallows on such skimpy evidence? As a piece of advocacy, it was superb, reducing a well-rounded prosecution case to a single scrap of disputed evidence. Once again, a jury rejected the saliva test and returned a not-guilty verdict.

Conclusion

That night at a hotel, Williams celebrated his freedom with Norman Rae, a newspaper reporter who had championed his innocence. In the middle of the night, Rae was awakened by Williams's pounding on his hotel door. Overcome with drunken remorse, the old man sobbed, "I've got to tell somebody. You see the jury was wrong . . . it was me."

Rae was appalled. But there was nothing he could do. Under the stringent British libel laws, publication of the confession could result in massive damages. And there was no point. Having once been found innocent, Williams could never again be tried for the same offense. For more than a decade, Rae kept news of the confession to himself. Only after Williams's death in 1951 did he reveal how he and the jury had been duped.

W. Thomas Zeigler Jr.

DATE: 1975

LOCATION: Winter Garden, Florida

SIGNIFICANCE: Bloodstain analysis originated with the great French criminologist Alexandre Lacassagne, but even he would have marveled at the expertise demonstrated in this case.

At 9:15 P.M. on Christmas Eve, 1975, the police in Winter Garden, a sleepy town in central Florida, received an emergency call from prominent local businessman Tommy Zeigler: He had been shot in an attempted robbery at his family's furniture store and needed help desperately. Patrol cars converged on the Tri-City Shopping Center. They found a bloodbath—four people shot and beaten to death. Only Tommy Zeigler was still alive, bleeding from a stomach wound. The dead included Zeigler's wife, Eunice, and her parents, Perry and Virginia Edwards. The other victim was a black citrus worker named Charlie Mays who, according to Zeigler, had been part of the gang that raided the store.

Zeigler responded well to treatment—his injuries were less serious than first thought—and he was able to give an account of the nightmarish events. Apparently he and Ed Williams, a black part-time employee, had returned to the store at just after 7:00 P.M. to meet Charlie Mays, who had a TV set on layaway that he wanted to collect for Christmas. Zeigler said that once inside the store, he was jumped by several attackers. In the melee he lost his spectacles, so he was unable to provide a good description of his assailants, other than that one was large and black. The brawl spilled over into Zeigler's office, where he had grabbed a .357 Magnum and had begun firing until he was hit by a bullet and collapsed.

Police Chief Don Ficke listened to this story thoughtfully. In recent weeks, two black desperadoes known as the Ski-Mask Bandits had terrorized local businesses. Besides robbing their victims, the two thugs often sexually assaulted them as well. Ficke, aware that Mays's pants had been undone, wondered if perhaps the dead man comprised one half of the notorious duo and Ed Williams the other. But something else troubled the police chief: By chance a stray bullet had hit the store clock and stopped it at 7:24 P.M. Why, Ficke mused, if the shootings had taken place before half-past seven, had Zeigler taken almost two hours to call for help?

Conflicting Story

The next day, Ed Williams went to the police of his own accord. He had heard the rumors and wanted to set the record straight. In his version of events, he and Zeigler met outside the store at around 7:30 P.M. as arranged. Zeigler had gone inside alone, then called out for Williams to follow. When he did so, Zeigler attempted to shoot him with a pistol, but the weapon jammed, allowing the terrified Williams to run for his life.

With such a welter of contradiction, obviously someone was lying, but much of the haze cleared when FBI ballistics experts reported that the bullets that killed Eunice Zeigler and her parents came from a pair of .38s known to have been purchased by Tommy Zeigler. Twenty-eight bullets were fired in the store that night, from eight different guns, all of them owned by Zeigler. Then came news that the Zeiglers had been having serious matrimonial trouble: Eunice had even talked of leaving Tommy after Christmas. His response was to insure Eunice's life for $520,000. It was a clear motive, enough to arrest him for murder.

Aware that Zeigler was someone of substantial means and therefore able to afford the very best in legal representation, state prosecutors knew they would have to present a watertight case in order to secure a conviction. As a consequence they asked Herbert MacDonell, an ex–industrial scientist who had turned the study and interpretation of bloodstains into a life's work, to examine the blood-drenched store. MacDonell's first view of the store came on January 7, 1976, and lasted all day. He spent hours on his hands and knees, measuring precisely, peering through a magnifying glass, comparing stains with detailed crime scene photographs. Underpinning the observation was cold, hard logic. The sum total enabled MacDonell to compile this chilling scenario:

Zeigler and his wife had entered the store through the rear door. Eunice went into the kitchen, where she was shot in the back of her head. Her casual attitude—no sign of disarray and her left hand still in her pocket—suggested to MacDonell that she had not seen any other bodies and was therefore the first person to be killed. A second bullet fired at Eunice had pierced the kitchen wall, stopping the clock on the other side of the wall, and timing the exact moment of her death at 7:24 P.M. MacDonell felt confident in saying this because the bullets, fired from the same weapon, had parallel trajectories.

Perry and Virginia Edwards were unintended victims. On their way to church, they had probably noticed lights in the store and gone to investigate. Instead, they blundered into a diabolical murder plot. Ten feet inside the rear door, Perry Edwards was struck by a bullet. Dripping blood, he lurched toward the kitchen. Zeigler went after him with a linoleum crank. Edwards had reached Eunice—evidenced by his blood dripping onto her dead body—when Zeigler shot him again. Virginia Edwards, cowering behind a sofa to escape her homicidal son-in-law, raised a protective hand as Zeigler loomed over her. A bullet grazed her finger before smashing into her skull.

Unwitting Accomplices

Now Zeigler waited for Charlie Mays and Ed Williams, unwittingly cast in the role of the Ski-Mask Bandits, to arrive. But the fates conspired once more against Zeigler—first the Edwardses, now Mays. The intended fall guy had company. Felton Thomas told investigators that he had gone with Mays to the store at about 7:30 P.M. Zeigler had met them outside with a strange request: there was a bag full of guns on his car seat; would they mind test-firing the weapons for him at a nearby orange grove? Although bemused, both men agreed and climbed into Ziegler's car. (Corroboration of this unlikely-sounding tale came when Thomas led officers to the orange grove and they found fragments of .38 caliber shells.) Afterward, Zeigler drove at breakneck pace back to the store. Once there, he announced that he had forgotten his keys. If Charlie wanted his TV set for Christmas, he said, they would have to break in. Thomas, not about to break into any white-owned store, refused and advised Mays to do likewise. After ostensibly driving home to get the keys, Zeigler entered the store with Mays. Still apprehensive, Thomas waited outside for a while. When Mays did not reappear, he left.

Inside the store, Mays was shot twice and then beaten to death with the linoleum crank. Although droplets of his blood fell onto that of Perry Edwards, the two had not mixed, an indication that Edwards had already been dead for some time. A shoulder holster, found near Mays's body and meant to incriminate him, was an obvious plant. The kind of beating that killed Charlie Mays would have sprayed medium-velocity bloodstains over the holster's upper surface. There were none. The back, too, was clean, even though the floor where it lay was drenched in blood. Clearly, the blood had dried—a process that MacDonell estimated would have taken at least fifteen minutes—before the holster was dropped onto it.

But Zeigler's own blood branded him a liar. He had been shot, he said, at the rear of the store, and from there he had struggled to the phone to summon help. Afterward he had waited at the front of the store. MacDonell tracked the drops of blood from the telephone to the store front, but could find none elsewhere. This meant that Zeigler had actually been standing by the phone when he was shot, and the only person who could have fired that shot was Zeigler himself.

On July 2, 1976, after a trial in which MacDonell's vivid reconstruction held the court spellbound, Zeigler was convicted of quadruple murder, and he was later sentenced to death.

Conclusion

In 1988 an appellate court decision overturned Zeigler's death sentence and he was awarded a new sentencing hearing. The following year, the original sentence was reimposed and Zeigler returned to death row. Always a notorious case, the Christmas Eve murders were featured on several TV programs, most of which portrayed Zeigler in a sympathetic light. These programs have been instrumental in the formation of a well-organized campaign to prove Zeigler's innocence. Among his more prominent supporters is the human rights activist Bianca Jagger. In April 2005 a motion for a new trial, based on DNA evidence that was unavailable in 1976, was rejected. At the time of this writing, Zeigler has been on Florida's death row for more than thirty years. Even so, this doesn't make him the Sunshine State's longest-serving condemned prisoner. Five other convicted killers stand between Zeigler and that dubious crown.

Arthur Hutchinson

DATE: 1983

LOCATION: Sheffield, England

SIGNIFICANCE: Several forensic specialties were employed to trap this sadistic killer.

One of the most sickening mass murders ever recorded in Britain erupted in the unlikeliest of settings. Basil and Avril Laitner were a wealthy professional couple with a large house in the fashionable Sheffield suburb of Dore. On October 23, 1983, their eldest daughter got married. As the Laitners entertained wedding guests in a large tent erected on the garden lawn, they were unaware that a stranger hidden in some nearby shrubbery was watching their every move. In particular, the voyeur was infatuated by Nicole Laitner, the eighteen-year-old sister of the bride.

Sometime in the early hours of the next morning, the stranger let himself into the house through a faulty patio door. Armed with an eight-inch bowie knife, he mounted the stairs. Nicole was awakened by the sounds of a violent commotion. The next moment her bedroom door flew open. A flashlight beam blinded her in bed. A man's voice warned her not to scream or he would kill her. He had already killed her parents, he said. At knifepoint, he led Nicole out onto the landing.

Dumb with terror, she picked her way through the blood that puddled around the lifeless body of her father. He had been sliced to ribbons. Her mother, too, had been butchered by twenty-six knife blows. In another bedroom, a third family member, Nicole's brother Richard,

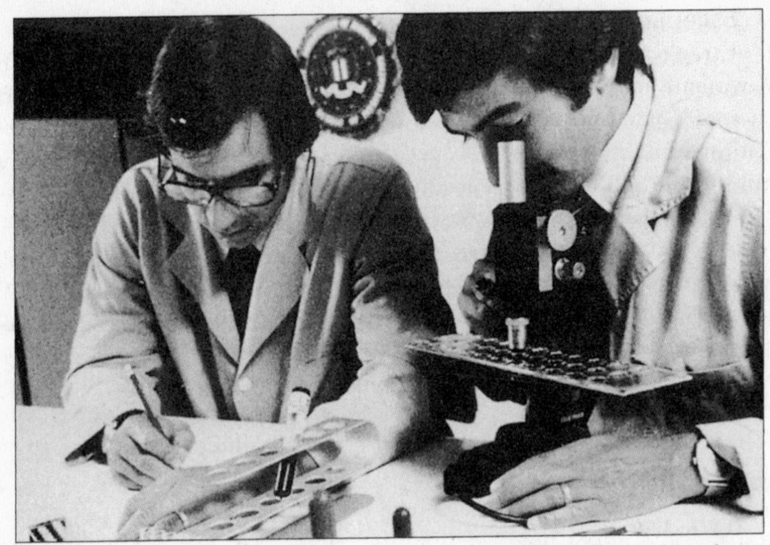

Serology unit examiners microscopically examine a liquid blood sample to determine ABO blood type. (FBI)

also lay dead. The killer—ill-kempt, stinking, and clad in a bloodstained T-shirt—steered Nicole through the carnage and out to the tent, where he made her strip naked. Sticking the knife into the ground by the side of her head, he growled, "You have got to enjoy it or I will kill you. That's where your mom went wrong. She created a fuss, so I had to kill her."

He later led the weeping girl back up to her bedroom, raped her twice more, tied her up, and made off. Nicole managed to free herself from the bonds the next morning at about the time some local workmen arrived to dismantle the tent.

Forensic scientists had rarely encountered so much blood at a crime scene; the landing, stairs, and two bedrooms were drenched. Two places where they unexpectedly found it, however, were on Nicole's bed, at around knee level, and on her nightdress. As she had not been cut during the attack, it was reasonable to assume that these bloodstains had come from the assailant. Alfred Faragher of the Home Office Forensic Science Service at Wetherby studied this sample carefully, identifying a rare combination of factors that occurred in only one fifty-thousandth of the population. And there was something else.

Escaped Rapist

Faragher had seen this same type of blood only a month previously. A woman had been raped in Selby, another Yorkshire town. The suspect, a forty-two-year-old petty thief and keep-fit fanatic named Arthur Hutchinson, had escaped from custody on September 28 by hurling himself through a second-floor window and scaling a twelve-foot-high wall topped with barbed wire. Hutchinson had a long record, including two convictions for sex offenses. His taste for violence surfaced early; at seven, he had stabbed his sister. Although most of his prison sentences were for minor offenses, he had recently been released after serving five years for carrying firearms and threatening his brother with a shotgun. (It was later learned that while in prison, Hutchinson filled two exercise books with notes about women who were likely candidates for rape and robbery, culling information from newspaper reports.)

With every police officer in northern England on his trail, Hutchinson checked into a hotel, signing the register "A. Fox." But Hutchinson, with his bloated impression of his own cleverness, was in for a shock. While scaling the barbed-wire fence, he had gashed his leg badly; now police let it be known that the barbed wire had been specially dipped and that any injury left untreated was likely to turn gangrenous. Hutchinson, who was known to frequent hospital emergency rooms for even the most minor ailment, took the bait. After being treated at Doncaster Royal Infirmary for a gashed knee, he was captured. Fittingly, the "Fox" was run to earth in a turnip field.

At his resulting trial, which began at Sheffield Crown Court on September 4, 1984, Hutchinson faced charges of triple murder. The evidence against him was overwhelming. Not only did Nicole Laitner identify him as her attacker, but he had left an impression of his shoe in blood on the stairs and a palm print on a bottle of champagne in the tent. Dr. Geoffrey Craig testified that two bite marks on a piece of cheese in the kitchen refrigerator matched Hutchinson's dental characteristics exactly. History was made when for the first time in a British murder trial, the jury was shown a police video of the murder scene, complete with bodies. The tape, lasting seven minutes, spoke more eloquently of the horrors at the Laitner household than could any advocate.

With so much evidence to the contrary, Hutchinson could hardly deny his presence at the house, but with vile cynicism he claimed he was there as Nicole Laitner's consensual sexual partner. All of the murders, he said, had occurred after his departure. Hutchinson's desperation reached its nadir when he suddenly gesticulated toward *Daily Mirror* crime reporter Michael Barron in the press box and yelled, "There's your killer!" It was a psychopathic act from a psychopathic defendant. He received three life terms.

Conclusion

Advances in serology now make it possible to identify some three hundred group systems, but the rush to develop a so-called blood fingerprint has been largely undermined by recent successes with DNA typing. Whether that particular scientific goal will ever be achieved remains to be seen.

TIME OF DEATH

Three traditional indicators are used to determine how long a person has been dead: rigor mortis, lividity (or hypostasis), and body temperature. Despite what detective novelists would have you believe, none is wholly reliable; all three can be hastened or slowed by a number of factors, such as ambient temperature, physique, exercise, alcohol, and drugs.

Rigor mortis usually begins to set in three hours after death in the muscles of the face and eyelids and then spreads slowly through the body to the arms and legs, taking about twelve hours to complete. In most cases the process reverses after thirty-six hours, until the body is soft and supple again.

Lividity also develops in a time sequence. It commences shortly after the heart stops mixing the plasma and red cells together and is visible thirty to sixty minutes later. As the red cells settle, like sediment in a wine bottle, the skin beneath turns red. In a process that takes between six and ten hours, the red cells break down, evacuate the capillaries, and enter the body. The color then becomes permanent. If a person dies in bed lying on his or her back, the color would normally be found on the back. Lividity on the front part of the body indicates that the body has been moved.

The third major indication of how much time has elapsed since death is body temperature. When oxygen is no longer fueling the body and keeping it warm, the temperature falls at a rate of approximately 1.5°F per hour. Again, musculature and ambient temperature play significant roles. An obese person will cool much more slowly than a thin person; someone who dies in a warm room will retain more body heat than someone who succumbs outdoors in cold weather. The following formula is widely used for estimating the time of death:

$$\frac{\text{normal temp (98.6 °F)} - \text{rectal temp.}}{1.5} = \text{approx. no. of hours since death}$$

None of these tests is 100 percent accurate; only by factoring in all the results can a likely time of death be established.

One recent addition to the medical examiner's armory, so-called eyeball chemistry, works by measuring the rate at which potassium from the red blood cells enters the vitreous fluid of the eye. Various studies of this technique have produced widely divergent results, and the jury is still out on whether this will ever enter the forensic medicine mainstream.

In 2001, scientists at Oak Ridge National Laboratory in Tennessee released details of a planned scanner that incorporates mass spectroscopy and gas chromatography and that can be passed over a corpse to detect and analyze the chemicals of decomposition. The belief is that these chemical "signatures" can be analyzed to give a time of death with only a two-day margin of error for every thirty days of decomposition. Research on this project is still continuing.

These are the conventional means of establishing time of death, but science is constantly coming up with improvements, as the following cases reflect.

Frank James and Raymond Schuck

DATE: 1920

LOCATION: Camden, New Jersey

SIGNIFICANCE: This case provides a brilliant example of deductive logic by the detective who was known as the Cornfield Sherlock Holmes.

Each week, David Paul, a sixty-year-old runner for the Broadway Trust Bank in Camden, New Jersey, carried a deposit across the Delaware River to the Girard Trust Company in Philadelphia. On October 5, 1920, with his satchel bulging with forty thousand dollars in cash and thirty thousand dollars in securities, he left Camden as usual but then disappeared. When a background check revealed that twenty-five years earlier Paul had served a jail term for mail theft, fears grew that he had absconded with the money. To their credit, his employers steadfastly refused to accept this view, insisting that Paul had reformed.

Eleven days later, their faith was vindicated in the cruelest way imaginable: Two duck hunters found Paul's body in a shallow grave near Irick's Crossing in adjoining Burlington County. He had been shot twice and battered about the head. No money was found on his body, but all of the securities were in the pouch. The killers had not bothered with trifles; Paul's watch, gold ring, and gold cuff links still lay on the body. Also close at hand was a pair of spectacles. Curiously, while the ground around the corpse was dry, Paul's overcoat and clothing were soaking wet. The thick clay soil showed faint traces of automobile tire tracks, and a piece of wood from the backseat of a car was found. Witnesses spoke of seeing a yellow runabout with a wooden rear seat in the vicinity several days before.

When the medical examiner declared that death had occurred not more than twenty-four hours before the body was discovered, it appeared that either Paul had actually bolted with the money, only to be killed later by accomplices, or he had been kidnapped and kept imprisoned for at least a week before being killed.

Neither theory made much sense to Ellis Parker, chief of detectives in Burlington County and one of the nation's top investigators, a man often consulted by other agencies when they faced an especially difficult case. Parker, who knew Paul well and had identified the body, was convinced that the messenger had been murdered much earlier than the medical examiner said. He was particularly intrigued by the wet clothing. The official line was that the murderers, unsure whether the gunshots had definitely killed Paul, had made certain by drowning his body in a nearby stream before burying him. Parker growled his disagreement and continued looking.

Two Suspects

Within a month, the spectacles were traced to a neighbor of Paul's, an automobile salesman named Frank James. It turned out that ever since Paul's disappearance, James and another man, Raymond Schuck, had been drinking wildly and tossing large sums of money around. Yet when questioned, both had watertight alibis for the time when the medical examiner estimated the murder had taken place: James had been at a convention in Detroit, and Schuck had gone downstate to stay with friends for several days. Parker, always wary of perfect alibis, especially when provided by people he believed to be crooked, redoubled his efforts.

Just upstream from where the body was found lay a number of tanning factories. Analysis of water from the stream revealed an unusually high tannic acid content, enough to act as a preservative on a human body, so that after a week or so in water it would show hardly any sign of decomposition. Submergence of Paul's body in the stream would account for the medical examiner's confusion regarding the time of death.

A triumphant Parker confronted James and Schuck with this latest finding. James gave it a moment's thought, then admitted everything. Schuck, who tried to pile all of the blame on his partner, admitted to dumping Paul's body under a bridge. After a mammoth drinking spree, they had returned to the bridge, retrieved the body, and buried it, unaware of the alibi that the water had given them.

A jury later apportioned equal guilt to both men, and they were electrocuted on August 30, 1921.

Conclusion

Over a career that spanned four decades, Ellis Parker solved 226 out of 236 murder cases, combining scientific detection with flashes of intuitive brilliance, all laced with prodigious determination. Often he would decide early in an investigation who the guilty party was and stubbornly pursue only that possibility. This tenacity served him well more often than not, but ended up ruining his career during the Lindbergh kidnapping, in which his meddling earned him a jail sentence. He died in prison in 1940.

Anibal Almodovar

DATE: 1942

LOCATION: New York, New York

SIGNIFICANCE: In this amazing case, forensic botany destroyed a phony alibi.

A New Yorker out exercising his German shepherd in Central Park happened to follow the barking dog into some tall grass. Beneath the low-hanging branches of a dogwood tree, Fridolph Trieman found the body of a young woman. She had been strangled. An examination found no evidence of rape, and the fact that she had no handbag or money suggested that this was a mugging gone tragically wrong—except that the woman still wore a gold chain around her swollen neck. No thief was likely to leave that.

Late that night detectives Joseph Hackett and John Crosby of the Missing Persons Bureau identified the woman as Louise Almodovar, a twenty-four-year-old waitress who lived with her parents in the Bronx. They had reported her missing the previous day, November 1, 1942. They told how Louise had married Anibal Almodovar, a diminutive Puerto Rican ex-sailor, just five months earlier, only to leave him after a few weeks because of his insatiable womanizing.

When told of his wife's fate, Almodovar just shrugged. She had made his life hell, he said, beating up one of his girlfriends and swearing at another. He was glad to be rid of her but denied any involvement in her death. And the facts seemed to bear him out. According to Thomas A. Gonzales, chief medical examiner, Louise had met her death between nine and ten o'clock on the night of November 1, at which time Almodovar was at a dance hall called the Rumba Palace with the very woman whom Louise had beaten. Furthermore, dozens of other witnesses could testify to his presence there. When faced with such an ironclad alibi, the detectives looked elsewhere. Then Louise's parents showed them threatening letters written by Almodovar to their daughter. It was enough to hold the amorous seaman as a material witness.

Still, there was that alibi. Only when detectives visited the dance hall, just a few hundred yards from the murder scene, did they realize that Almodovar might have slipped out unnoticed, gone to Central Park where he had previously arranged to meet his wife, killed her, and then returned to the dance without anyone being the wiser.

Seeds of Doubt

Forensic tests confirmed their suspicions. When Alexander O. Gettler, head of the city's Chemical and Toxicological Laboratory, examined crime scene photos, he noticed that the body was lying in an unusual type of grass. Enlargement of the photographs allowed him to identify the individual strain of grass. Coincidentally, grass seeds of the same type had been found in Almodovar's pockets and trouser cuffs, yet he insisted that he had not visited Central Park for over two years. Any seeds in his pockets, he said, must have been picked up on a recent visit to Tremont Park in the Bronx. He was wrong. Joseph J. Copeland, professor of botany and biology at City College, later testified that the grasses in question—*Plantago lanceolata, Panicum dichotimoflorum,* and *Eleusine indica*—were extremely rare and grew only at two spots on Long Island and three places in Westchester County. The only place in New York City where such grass occurred was Central Park. Moreover, it could be further isolated to the very hill where Louise's body had been found.

Almodovar panicked, suddenly recalling a walk he had taken in Central Park two months previously, in September. Copeland shook his head. The grass in question was a late bloomer, mid-October at the earliest; therefore Almodovar could not possibly have picked up the seeds in September. But on November 1?

At this, Almodovar broke down and confessed. Yes, he had arranged to meet his wife in Central Park; they had quarreled again, and he had killed her in a fit of rage. Later, he claimed that this confession had been extorted from him under pressure and pleaded innocent at his trial. But it did no good. When the sentence of death was passed, Almodovar, despite being shackled from head to toe, fought so fiercely that nine guards were needed to restrain him. On September 16, 1943, he died in the electric chair.

Conclusion

There can be little doubt that this was a carefully planned and executed killing, wholly at odds with Almodovar's claim of murder committed in the heat of passion. The alibi was just too good; fortunately, so was the scientific analysis.

Steven Truscott

DATE: 1959

LOCATION: Goderich, Canada

SIGNIFICANCE: A young girl was dead. Seven years later, pathologists from around the globe gathered in Canada to argue over when, exactly, she had been murdered.

The weather was fine on the evening of Tuesday, June 9, 1959, when Lynne Harper, twelve, left the Royal Canadian Air Force base where she lived at Goderich, near Clinton, Ontario. The time was half past five; she had just finished her supper and was now out to enjoy the summer sun. Soon afterward, she was seen with classmate Steven Truscott, fourteen, on the crossbar of his green racing bicycle, heading away from the base along a country road. Another child, cycling in the opposite direction, passed them at about 7 P.M. opposite a small wood called Lawson's Bush. Yet just a few moments later, when another boy came along, the road was deserted.

That night, Lynne failed to return home. A search lasting into the next day found nothing. On Thursday groups of both police and air force personnel fanned out across the fields and woods. That afternoon, Lynne's body, already starting to decompose in the hot weather, was found in Lawson's Bush, partly concealed by branches. She was lying on her back, seminaked, strangled with her own blouse. Between her feet two small mounds of soft earth, bearing the imprints of crepe-soled shoes, had been pushed up, in an attitude that suggested rape. Due to the condition of the body, this could not immediately be confirmed.

The autopsy was conducted by local forensic pathologist Dr. John Penistan under less-than-ideal conditions; the room on the air base was small and was lit only by a single electric bulb. Even so, he confirmed that Lynne had been strangled and raped. From an examination of the stomach contents—turkey, cranberries, and vegetables, a meal she had eaten on Tuesday evening—he estimated that she had died not later than 7:30 P.M.

At which time Steven Truscott was still out and unaccounted for.

Phantom Car

He had arrived home at 8:30 P.M. When questioned, he admitted being with Lynne but claimed to have dropped her off at Highway 8, where she got into a gray Chevrolet with yellow U.S. plates. A medical examination of Truscott found graze marks on either side of his penis consistent with forcible rape, though he maintained they were from a rash he'd had for several weeks. His leg had been gashed a day or two earlier, and his trousers—which he had since washed but not well enough to get rid of grass stains on the knees—were torn. Although he was known to wear a pair of crepe-soled shoes, they were never found.

Despite the absence of direct proof linking Truscott to Lynne Harper's murder, the circumstantial evidence ensured a murder conviction. Astonishingly, he was sentenced to death by hanging. Though no one believed for a moment that this sentence would ever be carried out on a fourteen-year-old boy in Canada, it galvanized public opinion. Eventually the sentence was commuted to life imprisonment. Because of his age, he was sent to the Ontario Training School for Boys.

In 1966, journalist Isabel Le Bourdais published *The Trial of Steven Truscott,* in which she claimed that he had been wrongly convicted. The book, although replete with inaccuracies—especially about medical aspects of the case—succeeded in renewing debate, and a special appeal was filed to the Supreme Court of Canada.

Across the Atlantic, the two most prominent pathologists in Britain, Professors Keith Simpson and Francis Camps, each read the book and reached polarized opinions. Camps believed an error had been made in establishing the time of death; Simpson thought not. Underlying this disagreement was a long-running feud between the two scientists; for years they had toiled in each other's shadow. Camps, a robust, truculent character, never missed an opportunity to vilify the man he regarded as his archrival. Simpson, the veteran of more than one hundred thousand autopsies, was less strident but unparalleled on the witness stand.

In October 1966, both pathologists flew to Canada to testify at the appeal. Lined up in Simpson's camp, besides Dr. Penistan, were Dr. Milton Helpern, New York's chief medical examiner, and Dr. Samuel Gerber from Ohio. Appearing for Truscott, in concert with Camps, was Dr. Charles Petty from Baltimore.

When Did She Die?

At issue was the vexatious question of when Lynne had actually died. Penistan had based his original estimate on the stomach contents. When a meal is eaten, there is considerable variation in the rate of digestion and speed of emptying the stomach. The type of food eaten makes a difference; fatty meals slow the process. Also, metabolic rates vary, and even the same person may have different rates at different times, depending on health and emotional state. Fear, fright, injury, or pain can hamper digestion. For example, after severe head trauma, food may stay in the stomach for several days, looking as fresh as when it was swallowed. However, generally, the average meal stays in the stomach for a couple of hours or so.

Allowing for the considerable volume and only partial digestion of the food in Lynne's stomach, and in the absence of anything to indicate ill health, Penistan decided that she had died no more than two hours after eating the meal at 5:30 P.M.

Nonsense, said Camps; it was impossible to draw any such conclusion. He was adamant that death could have taken place at any time between one and ten hours after eating, plenty of time for the girl to have been taken away, raped, and murdered in the mysterious gray Chevrolet—the existence of which was never demonstrated—and then dumped. Simpson, well used to Camps's often bizarre flights of fancy, responded coolly. He first pointed out that the murder scene at Lawson's Bush left little doubt about where the crime had taken place and then went on to make clear his support for the oft-maligned Penistan. Milton Helpern was equally emphatic in his belief that Penistan's conclusions had been correct.

After considering both sides, the Supreme Court held that there was no justification for a new trial, and the sentence was reaffirmed. Released in 1969, Truscott disappeared from public view. After reemerging in 2007, he once again appealed his conviction. At this time of writing, the court's decision is pending.

Conclusion

Probably no other trial ever attracted such a wealth of forensic talent and diversity of opinion. The argument and counterargument generated by a half dozen of the world's premier pathologists has firmly established this case in the annals of legal medicine.

William Jennings

DATE: 1962

LOCATION: Gomersal, England

SIGNIFICANCE: Forensic archaeology, a science used in the United States for more than a decade, was virtually unknown in Britain until the investigation of this tragedy.

On a bitterly cold winter morning in 1962, three-year-old Stephen Jennings vanished from the family home at Gomersal in West Yorkshire. That afternoon—December 12—his father, William Jennings, reported the boy missing to local police. Scores of officers, together with local villagers, combed the surrounding countryside for the toddler, but heavy snow that would make this winter the worst in living memory hampered all attempts to find the little boy. When the thaw came in March 1963, efforts were resumed, but after several fruitless weeks the search steadily tapered off.

From the outset, police suspected that Jennings, a twenty-five-year-old petty criminal, knew more about his son's disappearance than he was letting on (two women claimed to have seen him carrying a sack containing something bulky on the day of the disappearance). And there had been a family history of so-called accidents. In July 1962, little Stephen had been admitted to a local hospital with badly scalded feet; the burns were incurred while he and the father were alone in the house. No charges were pressed. Then, in September, the child had been found wandering through the village, half naked, without shoes, his face bruised and bloodied. All of these incidents were recalled during intense questioning of Jennings, but he stuck to his story that the boy had been abducted by gypsies, and in the absence of any clues to link him to any crime, there was little that police could do.

Ostracized by other villagers, Jennings continued to abuse his children. Just one month after Stephen's disappearance, another son was treated at the hospital for facial injuries, then a cracked femur. Finally, in 1965, Jennings and his wife, Eileen, were charged with child neglect and jailed for eighteen months. After their release, the couple separated and later divorced. Jennings moved away and remarried, starting a new life.

Twenty-Five Years Later

A generation passed, and memories of Stephen Jennings faded, until April 7, 1988, when a Gomersal man who had participated in the original search was out walking his dog. Suddenly the terrier began barking at something in the undergrowth. The man went to investigate. Beside a stone wall and less than a mile from the former Jennings home, he saw a tiny skull and some other bones. Instantly his mind raced back twenty-five years, and he ran to contact the police.

In order to establish how long the skeleton had been in situ, a team from the archaeological department at Bradford University excavated the grave. They found that the body had originally been laid on top of the turf, then concealed with stones. Later, the nearby stone wall had collapsed, covering the body further. Because nearly all of the skeletal bones were intact and in position, the team believed that the body had probably been wrapped in something like a sack, because an exposed body would have had its bones dispersed by wild animals or the weather. Although nothing remained of the sack or any clothing, which had rotted away, the team found a pair of leather sandals with the skeleton. In a good state of preservation, they matched exactly the description of Stephen's footwear on the day he had disappeared.

The remains were examined by Home Office pathologist Dr. Somasundram Siva. With the aid of an odontologist, he estimated that the body was of a boy three and four years old. A fracture to one wrist corresponded to an injury that Stephen Jennings was known to have received shortly before his death. The cause of death was plain. Eight ribs were fractured, consistent, Dr. Siva said, "with being punched . . . or kicked from behind."

After keeping his grim secret for a quarter of a century, William Jennings was stunned when detectives arrived at his home in Wolverhampton to take him back to Yorkshire for questioning. On the long car journey, he broke down and admitted to killing his son but declared it accidental. During his trial, experts testified that only extreme and prolonged violence could have produced such appalling injuries. Jennings's account of Stephen "falling down the stairs" did not fit the medical facts, and on May 23, 1989, he was convicted of murder and imprisoned for life.

Conclusion

The basic tenet of forensic archaeology is that when a body is buried, the soil layer is disturbed, and then often replaced in the wrong order. By studying such clues as minor ground undulations, investigators can establish a surprisingly accurate time frame.

TOXICOLOGY

It is a sobering thought that for most of recorded history, poisoners have been able to dispense their potions secure in the knowledge that they were impervious to detection. All that was needed was a modicum of care. If one enjoyed patronage, then even that little inconvenience could be overlooked. By the seventeenth century, professional poisoners were regularly engaged by the moneyed and occasionally by the royal families of Europe whenever domestic difficulties threatened to get out of hand.

Easily the most popular poison was arsenic. Not for nothing was it known as inheritance powder, precipitating, as it did, abrupt changes in fortune, not to mention the family tree. Of course, arsenic was never entirely reserved for the privileged; the working classes, too, knew of its efficacy. Lives lived in the grim shadow of plague and countless other ailments could be snuffed out in the twinkling of an eye with no one to say whether death had resulted from natural or unnatural causes. But all that was to change.

Around 1790, a chemist named Johann Metzger discovered that if substances containing arsenic were heated and a cold plate held over the vapors, a white layer of arsenious oxide would form on the plate. Although this "arsenic mirror" could prove that food had been doused with arsenic, it could not tell if a body had already absorbed arsenic.

This problem was solved by Dr. Valentine Rose of the Berlin Medical Faculty in 1806. He cut up a corpse's stomach and its contents and boiled them into a kind of stew. After filtering the stew to remove any remaining flesh, he treated the liquid with nitric acid. This converted any arsenic present into arsenic acid, which could then be subjected to Metzger's mirror in the usual way.

But by far the greatest toxicological leap forward came in 1836 when

James Marsh, a middle-aged London chemist, invented a means of detecting even the smallest quantity of arsenic. It was similar to Metzger's method, but instead of allowing the vapors to rise up to the cold metal plate—with most of the gases escaping into thin air—the whole process took place in a sealed U-shaped tube in which the vapors could exit only via a small nozzle. The suspect material was dropped onto a zinc plate covered with dilute sulfuric acid to produce hydrogen. Any arsine gas was then heated as it passed along a glass tube, condensing when it reached a cold part of the tube to form the arsenic mirror. In a refined form, the Marsh test is still used today.

How many people are murdered each year by poison is unknowable. Confirmed cases make up a minuscule percentage of the annual homicide rate. Data compiled by the FBI show that out of 15,517 homicides in 2000, just eight were poison victims. Given the extraordinary sophistication and range of modern detection techniques, one can only marvel that poisoners persist in their misguided attempts to fool the laboratory. But their advantage lies in the fact that before it can be detected, poison has to first be suspected; and while the symptoms remain so tricky to diagnose, it is safe to say that poisoning will remain with us.

The following is a list of the most common poisons and their typical symptoms:

Acids (nitric, hydrochloric, sulfuric)—burns around mouth, lips, nose
Aconite—numbness and tingling in extremities
Arsenic—acute, unexplained diarrhea
Atropine (belladonna)—dilated pupils
Carbolic acid—odor of disinfectant
Carbon monoxide—bright cherry-red skin
Cyanide—quick death, red skin, odor of almond
Metallic compounds—diarrhea, vomiting, abdominal pain
Nicotine—convulsions
Opiates—contracted pupils
Oxalic acid (phosphorous)—odor of garlic
Strychnine—convulsions, dark face and neck
Thallium—hair loss

Mary Blandy

DATE: 1751

LOCATION: Henley-on-Thames, England

SIGNIFICANCE: This case concluded with the first murder trial to feature toxicological testimony.

Despite bearing several of the qualities that Georgian men found desirable in a wife—she was sweet-natured, attractive, and rumored to bring a dowry of one thousand pounds (four thousand dollars)—Mary Blandy was, at twenty-six, still unmarried. Every fellow who set his sights on this young woman found himself deflected by her ambitious father, a prosperous lawyer with decided views on what he was looking for in a son-in-law. He wanted someone with wealth and position.

Captain William Cranstoun had plenty of the latter, being the son of a Scottish peer, but his pocketbook fell short of Francis Blandy's expectations, and there was the problem of the wife he already had back in Scotland. Never one to let such trivia jeopardize an opportunity for financial advancement, Cranstoun entirely neglected to mention either impediment when he took up residence with the Blandys in 1746 in their home in Henley, a picturesque village on the River Thames, west of London. The domestic arrangement delighted both the smitten Mary and her snobbish father.

For six months all went well. Then Cranstoun decided to dispense with his wife. He wrote her a letter saying that because his military prospects were being impaired by his marital status, would she mind jotting a line to the effect that she had never been his wife at all, merely a mistress? Mrs. Cranstoun clearly did mind and took her errant husband to court. The ensuing publicity infuriated Mr. Blandy, who for the first time realized that his future son-in-law was penniless. Cranstoun was banished not only from the house but from Mary's affections as well. Or so the outraged father thought.

The couple continued to meet in secret. They had an ally in Mary's mother, especially after Cranstoun had lent her forty pounds to repay a debt in London. When, in 1749, Mrs. Blandy contracted an illness and died, Cranstoun, under siege from creditors, pressed Mary for settlement of the loan. Without any capital of her own, she had to borrow money to repay him.

Thinking how much simpler life would be if he could get his hands on Mary's dowry, Cranstoun reapplied himself to the task of winning over Mr. Blandy. Perhaps, he said to Mary, a "magic potion" might improve the old man's disposition? Quite by chance, he knew of a Scottish herbalist with just such a nostrum. Having managed to worm his way back into the Blandy household in 1750, Cranstoun apparently slipped some of the powder into Blandy's tea. A quite remarkable transformation took place. Blandy became personable, almost benevolent. But whatever Cranstoun had used, its effects were temporary; the next morning, Francis Blandy was his usual truculent self. So Cranstoun suggested that Mary continue the treatment, and in April of the following year, he sent her some of the powder with instructions that she administer it to her father in small doses.

Violently Ill

At once, Blandy began to suffer acute nausea and stomach pains, and when a servant, Susan Gunnel, suspicious and puzzled by the old man's decline, tasted some of the gruel that Mary had prepared for him, she, too, became ill. In the bottom of the pan she found a gritty white powder, which she scraped up and showed to a neighbor, who in turn sent it to an apothecary. Because there was not yet a reliable test for arsenic, scientific analysis was impossible. Nevertheless, Gunnel took her suspicions to Mr. Blandy, warning that he was being poisoned by his own daughter. Blandy called Mary to his bedside and asked her if she had tampered with his food. Mary, panic-stricken, turned pale and bolted from the room.

Inexplicably, Blandy allowed Mary to continue preparing his food, a decision that did nothing to arrest his deterioration. On the contrary, he became much worse. When the cook saw Mary throwing letters and a white powder onto the kitchen fire, she waited until Mary had gone and then rescued the powder. But it was too late to save the stricken Blandy. On August 14, 1751, he slipped into a coma and died.

That night, Mary offered the footman five hundred pounds if he would help her escape to France. His refusal forced her to make a run for it the next morning. An angry crowd gave chase—by now the circumstances of Francis Blandy's death were common knowledge—and captured her. Cranstoun, told of his fiancée's predicament, fled to Europe.

Mary Blandy's trial at Oxford Assizes on March 3, 1752, was over in a single day. The main prosecution witnesses, four doctors, all agreed that the well-preserved nature of Mr. Blandy's internal organs suggested arsenical poisoning and that the white powder they had analyzed was arsenic. But the experiment they used was dangerously rudimentary: One of the doctors told of applying a red-hot iron to the powder and smelling the vapor. The odor, he said, was clearly that of arsenic. More damaging was the testimony of Susan Gunnel and the cook, both of whom testified to seeing Mary put strange powders in her father's food.

Mary did nothing to deny to deny the allegation, other than to insist that the potions were to improve her father's temperament. Then why, asked the prosecutor, had she rushed to destroy the powder once she was aware of being under suspicion? Mary had no answer. The jury thought they knew. After just five minutes of deliberation, they found her guilty of murder.

At nine o'clock on the morning of April 6, 1752, dressed in black bombazine, her hands bound with black ribbon, she mounted the gallows, still maintaining her innocence. Her final words were to the executioner: "Do not hang me too high, for the sake of decency."

While Cranstoun must be considered fortunate to have escaped Mary's fate, it was a fleeting reprieve. Just a few months later, he died in conditions of great poverty and considerable suffering in France.

Conclusion

By modern standards, the medical testimony in this trial smacks of quackery, but it should be remembered that if science is a struggle of intellect over ignorance, the struggle had to begin somewhere.

Charles Hall

DATE: 1871

LOCATION: Greenland

SIGNIFICANCE: For almost a century, the mysterious death of arctic explorer Charles Hall puzzled criminologists and seafarers alike. In 1968, forensic scientists finally answered the enduring question—was Captain Hall poisoned?

In 1871, President Ulysses S. Grant authorized an expedition in search of the North Pole. On July 3, financed by a fifty-thousand-dollar grant from Congress, the 387-ton steam tug *Polaris* set sail from New London, Connecticut, under the command of Charles Francis Hall, an explorer with considerable arctic experience. Right from the start, the mission was blighted. Hall clashed repeatedly with the chief scientist, Dr. Emil Bessels, a haughty twenty-four-year-old German who also served as the ship's physician.

By September, the *Polaris* had reached the northwestern coast of Greenland, and Hall decided to anchor for the winter in an inlet that he named Thank God Harbor, some five hundred miles south of the Pole. Bessels and fellow officer Sidney Budington were furious; both wanted to head south for safer waters.

On the afternoon of October 24, Hall returned from a scouting trip and called for hot coffee. After drinking just half a cup, he pushed the coffee aside, complaining of its sweet taste. Half an hour later, he was doubled over in agony and vomiting, afflictions that Dr. Bessels attributed to apoplexy. That evening, the assistant navigator, George Tyson, wrote in his diary: "Captain Hall is sick; it is strange, and he looked so well . . . this sickness came on immediately after drinking a cup of coffee."

The following day, Hall seemed to rally a little, but later he suffered further attacks, despite, as Tyson noted, Dr. Bessels giving him "frequent medication." Tyson also noted that when Bessels was not around, the sickly captain showed definite signs of improvement. By November 3, Hall had gone into serious decline, his mouth surrounded by sores, talking wildly about being poisoned, excoriating Bessels as "that little German dancing master." A subsequent inquiry learned that Bessels treated Hall with injections of a liquid, ostensibly quinine, distilled from "white crystals."

"How Do You Spell Murder?"

In his more lucid moments, Hall was seen scribbling notes in his private journals. (Locked in a wooden case, these were later thrown overboard when all unnecessary items were jettisoned to lighten the ship because it was in danger of sinking.) On one occasion, he turned to Budington and asked, "Tell me, Sidney, how do you spell murder?" Eventually, Hall rejected all medication prescribed by Bessels, convinced it was poisoned. When Budington approached the doctor with an offer to taste the medicine first and thereby allay the captain's fears, Bessels angrily refused.

Late in the evening of November 7, 1871, Hall lapsed into a final coma. The next morning he died. Crew members could not help noticing that Bessels, and his assistant Frederick Meyer, who had also fallen foul of Hall, took the bereavement with sunny indifference. Three days later, Hall was wrapped in the American flag, placed in a coffin, and buried on the frozen shores of Thank God Harbor.

He left behind him a hellish situation. Trapped by ice, the *Polaris* was slowly crushed to splinters. Thirty crew members set out to trek overland. Over the next eighteen months, they suffered every extreme of hunger and hardship until, after one of the most heroic sagas of survival in the history of exploration, they were rescued on April 30, 1873. That summer, they told their story to a Washington inquiry, which ruled that Captain Hall had died "from natural causes, viz. apoplexy."

For almost a century, doubts lingered. Finally, in August 1968, Professor Chauncey C. Loomis of Dartmouth College and pathologist Dr. Franklin Paddock of Lenox, Massachusetts, flew to Thank God Harbor, determined to settle the mystery once and for all. The site had been marked by a tablet of oak and brass. After a few minutes of digging, they uncovered the coffin in its shallow grave and pried off the lid. Body and coffin alike were frozen into the permafrost, and Paddock had to perform his autopsy squatting astride the hole.

The corpse was remarkably well preserved, covered by a thin veneer of ice from the waist down. Apart from empty eye sockets and a shriveled nose tip, the face was intact and the rust-red beard and hair had lost none of their coloring. Although moisture had converted the trunk and limbs into adipocere, the internal organs were intact.

In what must have been one of the most awkward autopsies ever performed, Paddock carefully removed sections from the skull and what remained of the brain and heart, together with samples from the hair, beard, and a fingernail. He also gathered soil samples from the grave area. This material was sent first to the Public Safety Laboratory in Boston and then to the Toronto Center of Forensic Sciences, where it was subjected to neutron activation analysis (see the John Vollman case on page 313).

Poison Found

Dr. A. K. Perkons, director of the Toronto laboratory, found widely varying amounts of arsenic in the fingernail. At the tip, the nail contained 24.6 parts per million (ppm) of arsenic; at the base, it registered 76.7. Assuming a normal growth rate of 0.7 millimeters a week, Perkons concluded that in the last two weeks of his life, Hall had received a massive dose of arsenic.

Because the soil near the grave contained high quantities of arsenic

(22.0 ppm), it was necessary to exclude the chance that some may have migrated from the soil to the body. Yet, said Perkons, "such migration would not explain the differentially increased arsenic in the sections of both hair and nails toward the root end." Every sample was washed before irradiation. Had arsenic come from the soil, it would have been distributed uniformly, yet "in neither fingernail nor hair was this the case." Taking into consideration the symptoms attending Hall's death, which Perkons considered to be of paramount importance, he felt that "arsenic poisoning is a fair diagnosis." Paddock agreed, saying that the results confirmed that Hall had ingested "tremendous doses of arsenic in the last two weeks of his life."

Conclusion

So was it murder, or was it suicide? Given what we know of Hall's nature and the circumstances of his death, there is little reason to suspect the latter, which leads inexorably to the question of who the killer was. Clearly, Emil Bessels emerges as the likeliest candidate; he made his loathing of Hall plain. But did that loathing lead him to murder? At this distance in time, the answer will forever remain conjecture.

An interesting anomaly is that Hall complained of the coffee being sweet. For centuries, poisoners have favored arsenic precisely because it has no taste; yet, according to Gleason, Gosselin, Hodge, and Smith in *Clinical Toxicology of Commercial Products,* this is by no means certain. They cite cases in which people ingesting arsenic have noticed a sweetish metallic aftertaste. In every other respect, the symptoms they quote for arsenical poisoning—vomiting, delirium, coma, paralysis, skin eruptions—parallel those that Hall exhibited.

Robert Buchanan

DATE: 1892

LOCATION: New York, New York

SIGNIFICANCE: Often overlooked in the history of criminology is the role played by American newspaper reporters, especially those in New York City in the late nineteenth and early twentieth centuries. This case highlights one of the best.

Dr. Robert Buchanan was a man in a hurry. Born in Nova Scotia, he moved to Scotland, qualified in Edinburgh, married in Nova Scotia, and moved to New York in 1886 to set up practice. By day, he was a respected medical man; by night, he caroused in low-life bars and bordellos, ordering drinks and women with equal frequency. This lifestyle drove his wife, Helen, to obtain a divorce and return to the sanctuary of Nova Scotia. Buchanan took the split well. For some time, he had been eyeing the beefy charms of Anna Sutherland, a wealthy brothel madam some twenty years his senior. On November 28, 1890, one day after she had changed her will in his favor, Anna wed her best client.

Unfortunately for Buchanan, Anna demanded that she be treated like a proper wife, insisting that he stay home at night with her, which wasn't what the ambitious doctor had in mind at all. Although he thoroughly approved of Anna's money, he found her vulgar and embarrassing, especially now that he was amassing a roster of well-to-do patients. Matters came to a head in early April 1892 with Buchanan's announcement that he intended to sail to Edinburgh—alone—to further his medical studies. Anna told him straight: either she went with him or she would cut him out of her will. Buchanan didn't hesitate; the passage was canceled forthwith. Within days, on April 22, Anna suddenly fell ill. Less than twenty-four hours later, she was dead. The physician who attended her in that final brief illness certified the cause of death as a brain hemorrhage. Buchanan was too busy counting his inheritance—fifty thousand dollars—even to attend her funeral.

And there the matter might have lain had not the redoubtable Ike White, a reporter for the *New York World,* happened to visit the coroner's office, looking to rustle up some good copy. While there, he overheard an elderly man demanding that an investigation be opened into Anna Sutherland's death. Buchanan, he thundered, was a scheming fiend who had murdered Anna for her money. The man, who introduced himself to White as Anna's ex-partner, admitted that his concerns were not wholly altruistic—Buchanan had cost him a share of Anna's fortune—but he was adamant that justice, not money, was fueling his campaign. He had already taken his suspicions to the police, but they had refused to heed the word of a pimp. White, less moralistic, put out a few tentative feelers and learned that Buchanan was nowhere to be found. A call to Nova Scotia revealed that he was back in Halifax. What White heard next stunned him—only three weeks after Anna's death, Buchanan had remarried his first wife, Helen!

Studied Fellow Killer

Smelling the makings of a sensational story, White tracked down other employees of the late brothel madam. They confirmed the general perception of Buchanan as a mountebank. One recalled something else: During the trial of Carlyle Harris,[1] Buchanan had followed the proceedings with great interest, frequently referring to the accused as a "bungling fool" and a "stupid amateur." He had boasted of knowing how to avoid the telltale pinpoint pupils but refused to elaborate.

White took this knowledge to the doctor who had signed Anna Sutherland's death certificate. He flatly refused to consider the possibility that she had died from anything other than brain hemorrhage. As for the suggestion of morphine poisoning, he singled out the Harris case as proof that pinpointing of the pupils was always present and that Anna had displayed no such symptom.

Showing the indefatigability that was to hallmark his career, White refused to accept the doctor's findings. At the back of his mind was the memory of a friend who had suffered from an eye ailment that necessitated frequent trips to the oculist. On each occasion, White recalled, his friend had returned with unnaturally dilated pupils, the result of being treated with atropine drops. Was it possible, the reporter wondered, that Buchanan had put a few drops of atropine in his wife's eyes just before death, in order to counteract the contraction of her pupils? White rushed to the home of the nurse who had tended Anna in her final days. Her recollection was clear: On several occasions she had seen Dr. Buchanan stooped over his ailing wife, dropping some medicine in her eyes.

This was all White needed to mount a major newspaper campaign, urging the New York coroner to issue an exhumation order for the body of Anna Sutherland. On May 22, her body was disinterred from Greenwood Cemetery and dispatched to the Carnegie Institute for autopsy. The results were unequivocal: death from an overdose of morphine. Professor Rudolph Witthaus, an eminent toxicologist, found that the body contained one-tenth of a grain of morphine in the remains, which he estimated was the residue of a fatal dose of five or six grains. He also recorded his belief that atropine would have disguised the distinctive contractions of the pupils. These results coincided with Buchanan's return to New York. On the basis of Witthaus's report, he was arrested and charged with murder.

---

1. In 1891, Carlyle Harris, a New York medical student, was charged with murdering his wife. At first, her death, too, had been attributed to a stroke, but pinpointing of the pupils had led investigators to detect a morphine overdose. Harris was convicted and sentenced to death.

Bizarre Experiment

His trial began on March 20, 1893. In a grotesque interlude, the prosecution sought to strengthen its hand by actually killing a cat in court with morphine, then dropping atropine in the poor creature's eyes. It was callous but effective. The pinpoint reaction was retarded. Not to be outdone, the defense countered with a full-scale attack designed to show that the color reaction tests used by Dr. Witthaus to identify the presence of morphine were not infallible. Witthaus watched in disbelief as Professor Victor C. Vaughan produced results that seemed to wholly undermine the integrity of the Pellagri test, previously thought to be the surest method of distinguishing between morphine and the deceptive alkaloids produced by decaying cadavers. In dual experiments, Vaughan obtained the distinctive red reaction not only when morphine was present but also when the drug was absent. The court gasped.

Had the defense decided to leave well enough alone, Buchanan would likely have been acquitted, but some in the New York legal community felt that in the Harris trial, the defendant's decision not to testify had militated heavily against him. Unwilling to repeat the mistake, Buchanan's lawyers put their client on the stand. It was a disaster. Buchanan had a carping manner that only became more exaggerated under cross-examination. The prosecution trapped him in so many lies and contradictions that all of the doubt created by the scientific dispute was entirely canceled out. Buchanan limped from the stand in tatters. Nevertheless, the jury still took more than twenty-eight hours to declare him guilty of murder.

While Buchanan languished on death row, his lawyers launched an appeal based on the tainted scientific evidence. But Witthaus was ready for them. He had taken his courtroom humiliation badly and set out to reestablish his reputation by proving that Vaughan's methodology had been flawed. He eventually tracked the source of the error to impurities in the chemicals used by the other scientists.

Buchanan's appeal failed, and on July 2, 1895, he took his seat in the same chair at Sing Sing that Carlyle Harris had occupied just two years earlier.

Conclusion

Modern toxicological methods are fortunately much more sensitive than those available to Buchanan's prosecutors, and a ruse such as using atropine drops would not go undetected for long in a modern laboratory. Also, new methods of color reaction (updated from those used by Dr. Witthaus), thin-layer chromatography, and gas chromatography can identify the presence of most poisons down to amounts as small as one five-thousandth of a grain.

Eva Rablen

DATE: 1929

LOCATION: Tuttletown, California

SIGNIFICANCE: A combination of brilliant reasoning and meticulous forensic analysis produced headlines all across the United States.

By any stretch of the imagination, Carroll Rablen was a considerate husband. Unable to appreciate music himself—a war wound had left him deaf—he willingly ferried his fun-loving young wife, Eva, to the local square dances in Tuttletown, California, so that she might enjoy herself. For the wounded veteran, this meant silently watching as Eva, attractive and outgoing, reveled in the attentions of her numerous dance partners.

April 29, 1929, was just such an occasion. Rablen remained outside in the car while Eva danced the night away inside the town's schoolhouse. Later on, as promised, Eva took Carroll a tray of coffee and sandwiches, weaving her way cautiously across the tumultuous floor. Near the door, she accidentally bumped into another woman. The dancer laughingly shrugged off Eva's apologies, saying that it was nothing, and the moment was forgotten. Outside, Carroll took the refreshments from his wife. He and Eva exchanged a few words as he ate and drank; then she returned to the hoedown. A few minutes later, the bluegrass fiddles were drowned out by a howling scream. Puzzled dancers ran outside to find Carroll writhing in agony on the floor of his car. Between convulsions he gasped that the coffee had a bitter taste. Before medical help could arrive, Carroll Rablen was dead.

A tearful Eva could offer little in the way of illumination to investigators probing the strange death, other than to dismiss suggestions that her notoriously moody husband might have committed suicide. Most residents of Tuttletown, as well as the medical examiner, believed that Rablen had died from natural causes. His autopsy revealed nothing untoward, and analysis of the organs was also negative. But one person refused to accept these findings.

It was no secret that Steve Rablen had always regarded his daughter-in-law as a gold digger, only after Carroll's money, although some felt that it was more a case of the old man resenting Eva's influence over his boy, a sense that she had stolen him away. Whatever the reason, Rablen continually pestered the police with his suspicions that Eva had poisoned Carroll for a three-thousand-dollar insurance payout. More to get the cantankerous old man off his back than anything else, Sheriff Dampacher reluctantly agreed to once again search the schoolhouse.

Poison Found

After an hour's fruitless rummaging, Dampacher noticed that beneath the steps was a dark space made accessible by a broken plank. He thrust in a speculative arm and a few seconds later withdrew a small bottle. On the label, in bold type, was written "STRYCHNINE." Dampacher read on and saw that it had been prepared by the Bigelow Drug Store in Tuolumne, half a dozen miles away.

When Dampacher called at Bigelow's, drugstore clerk Warren Sahey produced the poison register. There was only one recent transaction involving strychnine, a bottle sold just three days before Rablen's death to a woman calling herself Mrs. Joe Williams. She needed the poison to kill gophers, or so she had said. When Sahey later identified Eva Rablen as "Mrs. Williams," it was enough for Dampacher to arrest the widow and charge her with murder. Eva, screaming that she had been set up by her father-in-law, insisted that she was innocent.

In arresting Eva Rablen, Dampacher was taking a bold step; after all, the medical examiner had not found any poison in the victim's body. Aware that their entire case hinged on this single purchase, the authorities ordered a reexamination of the bodily organs. The original analyst had been a person of little experience; this time prosecutors enlisted the aid of California's premier forensic scientist, Dr. Edward O. Heinrich.

Prosecutors, not wanting to tip their hand, decided to keep Heinrich's intervention under wraps. At this point in his career, the "Wizard of Berkeley," as Heinrich was sometimes called, had achieved the kind of celebrity that gave defense lawyers fits. And Eva's team was already hard at work, rounding up witnesses who would testify to Rablen's manic-depressive state and repeated suicide threats. And they had also uncovered a woman ready to provide Eva with an alibi at the time the poison was purchased in Tuolumne.

Besides the stomach contents, Heinrich analyzed the poison bottle, some of Rablen's clothing, and other items, including the dead man's car. After days in the laboratory, he isolated strychnine in the dead man's stomach. Also, there were traces of poison on the car's upholstery and in the cup that Rablen had drunk from. Employing yet another talent—that of a handwriting expert—Heinrich studied the poison register signed in the name "Mrs. Joe Williams" and compared it with known examples of Eva Rablen's handwriting. A week's worth of painstaking comparison convinced him that Eva had penned the false signature.

And then Heinrich produced his masterstroke.

Brilliant Inspiration

He had heard about the crowded dance floor and Eva's circuitous journey across it. The thought struck him that she might have collided with someone and possibly have spilled coffee on them. Sheriff Dampacher respected Heinrich's reputation enough to realize that it was an avenue of inquiry worth pursuing. Everyone who attended the dance was asked to search their memories. One young woman, Alice Shea, distinctly remembered Eva bumping into her, because the collision had resulted in a few drops of coffee being spilled on her dress. Furthermore, she had not yet washed the dress. Would they like to see it? Immediately, the dress was dispatched to Heinrich for examination.

So many people wanted to attend the trial of Eva Rablen, whom the arrow points to, that it was held outdoors.

The coffee stains showed clear traces of strychnine, irrefutably linking Eva Rablen to the poison.

At a pretrial hearing, local interest ran so high that the judge decided to hold the proceedings in an open-air dance pavilion. Hundreds of spectators craned to hear Eva Rablen plead not guilty. The trial was set for June 10, 1929, and promised to be a spellbinder. From all accounts, Eva's lawyers were mounting a creditable defense. They had found Carroll Rablen's first wife, and she was ready to confirm her ex-husband's suicidal tendencies. Once, she claimed, he had spoken of throwing himself into the machinery at the lumber mill where he worked.

Just before the trial, news of Heinrich's intervention leaked out, and the word was that the master had solved the case. So pervasive was this rumor that Eva's defense team requested a special court session at which they announced that their client now wished to plead guilty. By this means, Eva Rablen avoided the death penalty and was sentenced to life imprisonment.

Conclusion

In a career spanning five decades, Edward Heinrich solved hundreds of cases. Today, many of his pronouncements, although often brilliant, would be deemed inadmissible in court, but that takes nothing away from his triumphs. In stature, he assumed a role somewhat akin to that of Sir Bernard Spilsbury, his transatlantic contemporary. Both were held in awe by the public, and both repeatedly demonstrated their ability to sway juries. In 1953, Heinrich died suddenly at seventy-two.

John Armstrong

DATE: 1955

LOCATION: Gosport, England

SIGNIFICANCE: Although advances in pharmacology bring tremendous societal benefits, progress often has its darker side as well.

Early on the morning of July 22, 1955, a twenty-six-year-old nurse named John Armstrong of Gosport, near Portsmouth, called the family doctor, Bernard Johnson, to say that his five-month-old son, Terence, was very ill. Dr. Johnson knew Armstrong and his wife Janet, nineteen, well. They seemed to be an ill-starred couple. Their oldest child, Stephen, had died in March 1954, and just two months later their two-year-old daughter, Pamela, had suffered a sudden illness but fortunately had recovered. Now this.

When Dr. Johnson arrived at the house, the baby was already dead. Although he did not suspect foul play, Johnson was unable to identify the cause of death and accordingly notified the coroner. The body, the baby's bottle, and a pillow he had vomited on the previous evening were sent for examination by local pathologist Dr. Harold Miller.

In the larynx, Dr. Miller found a shriveled red shell similar to the skin of a daphne berry, a highly toxic fruit. Further shells were recovered from the stomach. Miller placed the berry shell in a jar of formaldehyde and the rest of the reddish-colored stomach contents in another bottle and stored both containers in a refrigerator. He asked the coroner to check whether the child had been exposed to daphne berries. An officer who visited the Armstrongs was shocked by their seeming indifference to the tragedy; they appeared to be more interested in watching TV. Nevertheless, there was a daphne tree in the garden, and it was bearing fruit, and Armstrong confirmed that the baby carriage had stood under it.

All of this seemed to bear out Miller's theory—until he went back to the refrigerator. The single shell was gone. It had dissolved in the now red-colored formaldehyde. The shells in the other bottle had also dissolved, deepening the already reddish hue of the stomach contents. Had these been the shells of daphne berries, as he had originally thought, they would not have dissolved as they did. Puzzled by this baffling turn of events, Miller sent both flasks, along with the pillow and feeding bottle, to a local laboratory that regularly performed toxicological investigations for the coroner. Their report did not include any mention of poison or traces of the daphne berry skin; the only unusual features were a small quantity of cornstarch and a red dye, eosin.

Poison?

Miller read this with some apprehension; such a combination was likeliest to occur in the red gelatin capsules that contained Seconal, a powerful barbiturate. This could mean deliberate poisoning. Already rumors had surfaced regarding the Armstrongs' perilous financial state. Was it conceivable that two parents would deliberately kill their own baby just to ease the family's financial burden? While no precedent existed for murder by barbiturates, Miller knew that a few grains of Seconal would certainly kill an infant. To satisfy his own curiosity, he dissolved three Seconal capsules in gastric juices. As anticipated, they reduced to cornstarch and eosin.

At this juncture, Miller passed the pillow and the other specimens to the police for more thorough analysis. Superintendent L. C. Nickolls, director of Scotland Yard's Forensic Laboratory, was assigned to the case. In the past, Nickolls had often expressed concerns about the burgeoning numbers of potentially lethal drugs—especially the soporifics—that were now in almost every home. In his experience, only the most sophisticated facilities were capable of detecting them. Nickolls's own laboratory was state of the art; even so, it still took five days of rigorous testing to establish that vomit traces on the pillow contained one-fiftieth of a grain of Seconal. He also isolated one-third of a grain in the stomach contents. When detectives called at the naval hospital where Armstrong worked and asked if any Seconal was missing, a nurse recalled the mysterious disappearance of fifty Seconal capsules from a cupboard to which Armstrong had access.

Meanwhile, inquiries began into the circumstances of Stephen's death the previous year, especially when it was discovered that the certificate had been signed by an eighty-two-year-old doctor who had never previously treated the child. The symptoms mirrored those suffered by Terence—drowsiness, blue-tinged face, breathing difficulty, and death. Two-year-old Pamela's sudden illness had taken the same course and might have reached a similar conclusion had it not been for prompt medical attention.

Nickolls's report concluded that the amount of Seconal extracted from the baby's organs indicated an original dose of between three and five capsules, absolutely lethal for a child that age. Concerns that they might be dealing with a serial killer led the police to exhume Stephen's body. John Armstrong watched the process without any undue distress. His only comment, "There won't be much left of him by this time, will there?" proved frustratingly accurate. Decomposition had destroyed all traces of any relevant chemical in the remains.

One Killer or Two?

All those who interviewed the Armstrongs felt they had planned the killing jointly. To clarify possible discrepancies in alibis, Nickolls needed to know how long a Seconal capsule of the type administered to baby Terence would take to burst in the stomach. He found that as the methyl cellulose capsule absorbed stomach fluids, it caused the cornstarch inside to swell. This in turn caused the capsule to erupt and discharge its contents into the stomach. Sometimes they opened swiftly, but it could take as long as ninety minutes.

This was of little help to those investigating the Armstrongs. Without any firm evidence to suggest that they were in possession of Seconal on the day of the murder, no charges could be brought against them. All they could do was wait.

A year passed. On July 24, 1956, Janet Armstrong applied for a separation and maintenance order against her husband, citing his repeated physical abuse. When the order was refused, officers lent a sympathetic ear to the disappointed plaintiff. Clearly embittered by the experience, she now turned viciously on her husband. Yes, he had brought Seconal capsules home from work, and they were in the house on the day of the murder. Three days after the baby's death, he had ordered her to throw all the capsules away. She claimed to have asked him, "Did you give the baby any?" to which he replied, "How do I know you haven't?" Fearful of his retribution, she had kept silent. Until now.

Janet Armstrong's bilious denunciation did nothing to dampen official suspicions of her complicity, and four months later she and her husband stood side by side in court, jointly charged with murder. What degenerated into an acrimonious verbal battle between the two ended with a split verdict. John Armstrong was sentenced to death; his wife walked away from the court a free woman. One month later, and beyond the reach of the law, she admitted giving Terence a single Seconal capsule. By this time, her husband's sentence had been commuted to life imprisonment.

Conclusion

The fact that Seconal had gone unrecognized in the local toxicological investigation—and would very likely have escaped notice in most other laboratories—emphasized the need for toxicological vigilance everywhere, because one never knew where the next poison was coming from.

Georgi Markov

DATE: 1978

LOCATION: London, England

SIGNIFICANCE: Everything pointed to murder, but proving foul play would lead scientists into the murky world of international espionage.

After defecting from Bulgaria in 1971, writer Georgi Markov took his antigovernment stance to London and began broadcasting programs to his homeland via the BBC World Service. His blend of satire and criticism infuriated Bulgaria's communist regime, and it was no secret that certain government departments desperately wanted him silenced for good.

On Thursday, September 7, 1978, their wish was realized. As the forty-nine-year-old Markov was waiting for a bus on London's Waterloo Bridge, he felt a sharp jab in his right thigh. He turned quickly to see a man behind him, brandishing a furled umbrella. The stranger mumbled an apology in a thick accent, then hurried off to catch a taxi. Markov thought little of it until reaching home that night. He told his wife of the mysterious encounter and then rolled up his trouser leg to show her a red puncture mark on his thigh.

The next morning he was admitted to a hospital with a high fever and vomiting. An X-ray of his thigh failed to reveal anything, although an area of circular inflammation now surrounded the puncture wound. Both temperature and blood pressure plummeted, while his pulse soared to 160 beats per minute. To their astonishment, hospital staff found that Markov's white blood count had risen to more than three times the normal level. Septicemia was diagnosed. Over the weekend, the patient became violent and confused. Massive amounts of antibiotics failed to arrest the decline, and on the following Monday he died.

Following an autopsy, the section of skin containing the puncture wound was sent to the government's top secret Chemical Defense Establishment at Porton Down, where Dr. David Gall, one of the country's leading experts on poisons and nerve agents, led the investigation. While examining the tissue, he found what first appeared to be a pinhead but later turned out to be a minute metal pellet, 1.5 millimeters in diameter, skillfully engineered, with two tiny holes drilled at right angles to each other. One of the holes was clear; the other was clogged with congealed tissue. Thoroughly bemused, Gall sent it to Dr. Ray Williams at the

Metropolitan Police Laboratory in London for further tests, while he got on with the job of identifying what had killed Markov.

Poison Pellet

Under a scanning electron microscope, the pellet was found to be made of 90 percent platinum and 10 percent iridium, an alloy stronger than steel and impervious to corrosion, and with the additional bonus of being radiopaque, almost invisible on an X-ray. Experts could think only that the pellet had been designed to carry a tiny amount of poison—no more than 2 milligrams—and that the lethal ball had then been fired into Markov's thigh by some kind of gas gun hidden in an umbrella. No other conclusion made sense.

And yet Gall's team was unable to detect any trace of poison in the body, which left them only with Markov's symptoms. What substance, they asked, would produce those kind of deleterious effects on a healthy and husky six-foot male? They soon rejected bacterial and viral infections, as well as most chemical poisons, because the amounts needed to cause death were greater than the pellet could have administered. One by one, the various toxins were eliminated, until they were left with a single substance that not only would induce all the right symptoms, but was sufficiently lethal to be transported in the minuscule metal ball.

Ricin, a derivative of the castor bean, is roughly five hundred times as toxic as arsenic or cyanide. It causes red blood cells to agglutinate before going on to attack other body cells with devastating effect. Death comes from an electrolyte imbalance. In order to test their theory, scientists at Porton Down injected a live pig with the amount of ricin the pellet could have contained. The unfortunate beast died within twenty-four hours. When examined, its organs revealed virtually identical damage to that found in the body of Georgi Markov.

Although compelling, the evidence was still circumstantial—for no trace of ricin itself could be found in Markov's body. Yet that factor alone provided even more reason to suspect that ricin was the culpable agent. Eventually, the body's natural protein-making cells break ricin down, so that having done its damage, the toxin disappears without a trace.

Having succeeded in finding out what had killed Markov, investigators fared less well in uncovering whose hand was behind the umbrella. Diplomats at the Bulgarian Embassy in London described allegations that their secret police had been involved as "absurd." But the evidence indicated otherwise. Just one year earlier, in an incident outside a metro station in Paris, another Bulgarian émigré, Vladimir Kostov, had been mysteriously jabbed, had then fallen sick, but subsequently had recovered. After some persuasion, he was visited by a surgeon, who extracted a small metallic object from his flesh. This was taken to the Forensic Laboratory in London, eagerly awaited by Ray Williams. The pellet was identical to the one that killed Markov. By chance, Kostov's pellet had lodged in muscle in his upper back, away from major blood vessels, and he had lived.

Conclusion

Not until 1991 and the collapse of communism did Bulgaria's new government reluctantly admit that their predecessors had carried out several assassinations of Bulgarian defectors, including Georgi Markov. A slow drip of information over the next decade revealed that the Bulgarian intelligence service (KDS) had actually sought permission from Moscow to eliminate the troublesome Markov. Not only had the Soviet KGB sanctioned this request, but they had also supplied the ricin. The name most often mentioned as the probable assassin is that of Francesco Guillino, a Dane of Italian extraction, who had reputedly worked for the KDS since 1972. Under the code name *Piccadilly,* this former drug smuggler made regular visits to London, and records show that one day after the attack on Markov he flew out of Britain for Italy. Guillino was interviewed in connection with the Markov killing as early as 1993, only to be released for a lack of evidence. The last reported sighting of him was in Copenhagen in 2004. Under pressure from both Western and homegrown journalists, on May 2, 2006, the Bulgarian Interior Ministry announced that it had started a procedure to open millions of secret files, creating the possibility that the mystery of who killed Georgi Markov might be cleared up once and for all. They'd better hurry, because in Bulgaria the statute of limitations for murder expires after thirty years. After September 2008, it will all become academic.

[illegible]

scanning electron microscope required [illegible] of research before [illegible]
coming fully operational. It works by scanning a sample, not with light, [illegible]
[illegible] electron beam [illegible]

TRACE EVIDENCE

"Every contact leaves a trace." So said the great French criminologist Dr. Edmond Locard in 1910, and these few deceptively simple-sounding words laid the cornerstone of modern forensic science. Put another way, every time a crime occurs involving physical contact, the perpetrator either leaves something at the scene or takes something away, often both. It might be hair, fibers, grit, powder, flakes of skin, a button, earth, or any of countless other items, all of which make the crime scene officer's job critically important, because evidence once missed is often irretrievable. By its very nature, trace evidence is often easy to overlook, and only the most diligent and painstaking examination will uncover all there is to be found, even if this means vacuuming the entire crime scene. Samples, when found, are logged and labeled, then passed to the laboratory for analysis.

Of course, there is little point in gathering this evidence—which may be barely visible or even invisible to the naked eye—unless it can afterward be seen clearly. For this reason, the handheld magnifying glass, although forever linked with images of Sherlock Holmes, remains the single most useful piece of scientific equipment available to an investigator, both at the crime scene and during initial laboratory examination. From there it is a short step to the compound microscope, invented between 1590 and 1609 by Dutch lens grinders.

Although a simple reflected-light microscope can enlarge samples by a factor of one thousand, it was found that still greater magnification was needed. First suggested by French physicist Louis de Broglie in 1924, the scanning electron microscope required a decade of research before becoming fully operational. It works by scanning a sample, not with light, but with an extremely fine electron beam, or microprobe, that generates electron emissions that feed back information about a sample's contours. After being amplified by a photomultiplier, this information can be displayed as three-dimensional enlargements of up to 150,000 times on a visual display unit, although most forensic work rarely requires magnification of more than ten thousand times.

The technique is particularly useful in comparing evidence such as paint fragments, fibers, paper, and wood and is capable of working with samples as small as one hundred-thousandth of an inch. Additionally, the scanning electron microscope can provide high-resolution photomicrographs for use as court evidence.

Another problem faced by laboratory technicians is identifying what exactly a sample is made of. Here the most significant breakthrough occurred in 1859, when two German scientists, Robert Wilhelm Bunsen and Gustav Kirchoff, discovered the principle of spectrography. They found that the spectrum of every element has its own individual signature, and passing light through a substance produces this spectrum, which the spectroscope reveals in a series of dark lines called absorption lines. By combining this with a related instrument called a spectrophotometer, which measures the intensities of light of different wavelengths, scientists can analyze and identify all manner of different substances.

Of course, the full array of techniques and instrumentation available to the modern forensic analyst is too vast to be covered here. Each decade brings inventions and refinements that narrow the gap between criminal and pursuer.

William Dorr

DATE: 1912

LOCATION: Lynn, Massachusetts

SIGNIFICANCE: Microscopy was unknown to most Americans until they read about this sensational case.

Millionaire George Marsh didn't appear to have an enemy in the world, yet someone pumped four bullets into the seventy-seven-year-old retired soap manufacturer, then dumped his body beneath an embankment abutting the salt flats near Lynn, Massachusetts. The crime seemed motiveless, because the victim's wallet and gold watch were untouched, and there was no likelihood of any family member benefiting by his death; Marsh had bequeathed his entire estate to charity, with the full knowledge and approval of his relatives, all of whom were independently wealthy.

He was last seen alive on the afternoon of April 11, 1912, and according to the medical examiner, he met his death sometime that evening. Clues were in short supply, apart from a scrap of woven cloth with a pearl-gray overcoat button attached, found a few yards from the body. It might have come from the killer; it might not. Only by finding the overcoat could investigators gauge its true significance.

Appeals to the public for information led to reports of a light-blue car seen near Marsh's home on the day of the murder. A boarding-house keeper recalled letting a room to a young man who drove such a vehicle. He had given his name as Willis A. Dow and spent considerable time at his window with a pair of binoculars, studying the rear of George Marsh's house.

Another landlady was even more helpful. She, too, had let a room to Willis A. Dow at the time of the murder, and on his departure he had left behind an overcoat from which all the buttons had been removed. Owing to a lack of local forensic facilities, police forwarded the coat and button to the Lowell Textile School in northern Massachusetts for examination.

Professor Edward H. Baker and Louis A. Olney compared both items under the microscope. The fragment of multicolored cloth on the button matched the cloth of the overcoat in both weave and texture, and the area of the coat from which it had been torn—with its broken threads—was also clearly visible. They speculated that, as he fought for his life, George Marsh had yanked the button from the coat of his killer. Later, the murderer had removed all the remaining buttons and presumably destroyed them in a futile attempt to thwart any subsequent identification.

Murder Weapon

While these findings inexorably linked the button from the crime scene to the overcoat, they in no way helped identify the mysterious killer. For that, detectives had to thank a fisherman who trawled a .32 Colt pistol from the water less than fifty yards from where Marsh had been killed. The manufacturer's serial number records led to a store in Stockton, California, and in turn to a thirty-year-old motorcycle dealer named William A. Dorr.

At this point, the saga took an incestuous twist. Dorr's aunt, Orpha Marsh, who also lived in Stockton, happened to be the foster daughter of George Marsh's deceased brother, James. As such, she was the heiress to James's considerable fortune. Dorr, a grasping psychopath, was not the kind of man to let a few bloodlines get in the way of some hefty capital gains, and he set about seducing his matronly aunt. He met with spectacular success. Orpha, thirty-eight years old, unmarried and unloved, was so smitten with her nephew's amorous overtures that she drew up a will naming him as sole beneficiary. The only problem with this arrangement, to Dorr's way of thinking, was that Orpha's fortune was held in a trust administered by George Marsh, who allowed her only a monthly stipend of $87.55. Dorr seethed; as long as the elderly millionaire remained alive, he would never get his hands on the capital. Over many months he convinced the ever-more-compliant Orpha that her affairs were being grossly mishandled, until finally she agreed that he should travel across country to confront her skinflint uncle.

Whether Orpha Marsh was privy to Dorr's murderous scheme remains a matter for conjecture. What is certain is that she received a letter from him informing her that the deed was done and enjoining her not to breathe a word about it until the matter was public knowledge—advice she took to heart.

But already the police, aware of her liaison with Dorr, had tapped her telephone, and they were listening when Dorr called her from a nearby restaurant. Dorr still had the phone in his hand when he was placed under arrest. He admitted killing Marsh but claimed to have acted only in self-defense because the near-octogenarian had attacked him! Afterward, he had propped the body up in the passenger seat of his car and driven around for thirty minutes before dumping it on the flats where it was found the next day.

Orpha, perhaps sensing her own precarious position, sealed Dorr's doom by releasing a diary in which he had detailed his daily stalking of George Marsh and hiring "a negro who will do the job for me for $1,000." This imaginary account seems to have been written purely for Orpha's benefit, perhaps an attempt by Dorr to convince her that his wasn't the murdering hand.

At his trial, Dorr's plea of mitigation received scant consideration from the jury, who decided his guilt after only two hours of deliberation. On March 24, 1914, he was executed in Charlestown, Massachusetts.

Conclusion

Millions of Americans read and marveled over the fact that a single overcoat button sent Dorr to the electric chair, as what had once been the preserve of the fictional detective now entered the realm of reality.

Colin Ross

DATE: 1921
LOCATION: Melbourne, Australia
SIGNIFICANCE: This was the first Australian murder case to obtain a conviction through forensic science.

Gun Alley, a shadowy cul-de-sac in the Eastern Arcade district of Melbourne, had long served as a favorite assignation point for prostitutes and their clients. For this reason it was regularly patrolled by the local constabulary. On the last day of 1921, an officer shone his light into the gloom, half expecting to find the usual furtive couples. Instead there was only a large bundle lying on the ground. He edged closer.

A young girl had been strangled and bludgeoned to death; her naked body, bruised but oddly free from blood, had been wrapped in a blanket and dumped like so much garbage. The savage nature of the rape confirmed that this was no ordinary killer. The victim was soon identified as Alma Tirtschke, a twelve-year-old schoolgirl who had been reported missing the previous day. Outraged by the crime's brutality, local prostitutes who had used Gun Alley on the night of the murder helped police compile a timetable that established that the body must have been dumped after 1 A.M.

As is often the case, the negative evidence spoke loudest of all. An absence of blood at the crime scene meant that the girl had been killed elsewhere and then transported to Gun Alley. Unless the killer had a car—rare in Australia at this time—the mission must have been undertaken on foot, implying that the killer lived close by. Rather more puzzling, the victim's body had been carefully washed and dried.

As the investigation gained momentum, one name kept cropping up—Colin Ross. Numerous witnesses placed this local wine-bar owner in Gun Alley at around the time the body was thought to have been dumped, including one person who claimed that he had been carrying a blanket-wrapped bundle. When questioned, Ross almost fell over himself to be helpful. In so doing, he talked himself to the gallows. Not only did he admit that Alma had been in his bar on the fateful day—though his claim to have scarcely noticed her hardly squared with his precise description of what she had been wearing—he also revealed details about the murder that only the killer could have known. Gradually, Ross's unsavory predilection for young girls came to light, a craving he himself had once summed up thus: "I prefer them without feathers."

Destroyed Evidence

Police theorized that Ross had lured Alma to his bar, where he assaulted and strangled her, then washed the body, presumably to destroy all traces of evidence. After wrapping the corpse in a blanket, he had made his way to Gun Alley, intending to hide it under a drain grating, only to be disturbed by the approaching police officer. In a panic he had abandoned the body and run off into the darkness.

Two blankets found at Ross's home were sent to Dr. Charles Taylor, a government industrial chemist, for analysis. On one he found several strands of reddish-gold hair, the same color as the victim's. His first task was to establish that the hairs were human; this was easily achieved as only the hair of some apes in any way resembles that of humans. The high degree of pigmentation indicated an age range of twelve to thirty, when coloring is at its peak; the length—some strands were more than twelve inches long—at that time virtually guaranteed that the hairs had a female origin. Some were pulled out by the roots, a common feature of assaults; others were broken off.

At Ross's trial the defense produced a fair-haired female customer from Ross's bar and challenged Taylor to distinguish between samples of her hair and that of the murder victim. He did just that, demonstrating clear dissimilarities between the two and proving, beyond a reasonable doubt, that Alma Tirtschke had been in contact with the blanket found at Ross's home. Despite his protestations of a police frame-up, Ross was found guilty and was hanged on April 24, 1922.

Conclusion

Even though forensic science was still relatively new to Australia, Ross's postmurder precautions provided disturbing evidence that killers were becoming increasingly aware of the need to cover their tracks. Fortunately, he didn't realize that murder victims—even clean ones—leave clues wherever they go. But not everyone is happy with Ross's conviction. Following the original publication of this book, relatives of Ross contacted the author with news that a reexamination of a single strand of hair from Alma Tirtschke had taken place. Although it was not possible to obtain the hoped-for DNA profile, it is claimed that modern comparisons of this hair with another hair taken from the crime scene show a clear difference. The authorities remain unconvinced. Thus far, they have rejected every attempt to have Ross's conviction overturned.

The D'Autremont Brothers

DATE: 1923

LOCATION: Ashland, Oregon

SIGNIFICANCE: Over a lengthy career, Edward O. Heinrich was credited with solving more than two thousand crimes; many consider this his finest hour.

America's last Wild West–style train robbery occurred on October 11, 1923, when two armed desperadoes clambered across the tender of a Southern Pacific express train just as it was about to exit a tunnel in the Siskiyou Mountains of southern Oregon. They ordered engineer Sydney Bates and fireman Marvin Seng to halt the train so that its engine, tender, and mail car were clear of the tunnel, leaving the rest of the coaches, those filled with passengers, in the blackness. A third gang member, waiting outside the tunnel, attached dynamite to the side of the mail car and then ran toward a detonator. Seconds later, an enormous explosion engulfed the mail car and its contents. Everything, including the lone mail clerk, Edwin Daugherty, was incinerated. Bates was then ordered to shift the train out of the tunnel. He did his best, but the big locomotive would not budge. His reward for failing to comply was a bullet. Fireman Seng and brakeman Charles Johnson, who had come forward to investigate the explosion, were also shot down in cold blood before the panicky killers made good their escape, empty-handed.

The attempted train robbery, so reminiscent of the James Gang, attracted virtually every available investigative agency: railroad police, postal detectives, and sheriff's deputies, as well as officers from nearby Ashland. A huge posse set out in pursuit of the bandits, but all they found was the battery-operated detonator, a revolver, a pair of greasy blue denim overalls, and some shoe covers made of gunnysack soaked in creosote, apparently to foil any tracker dogs.

As days and weeks passed with no discernible leads, someone suggested contacting Edward Heinrich. It was not a universally popular decision. Many doubted this quiet chemist from Wisconsin who had filled his crime research laboratory at Berkeley with the latest in scientific technology. Eventually, Heinrich was given the overalls, together with news that a garage mechanic had been arrested because his dungarees appeared to be stained with the same grease.

Heinrich examined every stitch of the overalls under a microscope, including samples of the grease and debris from the pockets. His first findings led to the immediate release of the luckless mechanic—the stain was not auto grease but pitch from fir trees. Then the scientist went on to amaze everyone by providing a full description of the man they should be seeking: a left-handed lumberjack from the logging camps of the Pacific Northwest, someone with light-brown hair who rolled his own cigarettes and was fussy about his appearance. He was five feet ten, weighed about 165 pounds, and was in his early twenties.

Incredible Deduction

How, listeners gasped, had he gleaned all of this from one pair of overalls? Heinrich was matter-of-fact. Because the pockets of the left side were more heavily worn than those on the right, and because the garment buttoned from the left, it was reasonable to assume that the wearer was left-handed. Chips of Douglas fir, common to the forests of the Pacific Northwest, were found in the right pocket, such as might be collected by a left-handed lumberjack standing with that side nearest the tree. Shreds of loose tobacco in both pockets were an obvious indicator of smoking preference.

Simple measurement of the overalls gave Heinrich a good idea of the owner's build and height, while a pocket seam yielded several neatly cut fingernail parings, somewhat incongruous for a lumberjack unless he was fastidious about his appearance. A single strand of hair clinging to one button was light brown; the level of pigmentation suggested someone in his early twenties.

In addition, Heinrich found one other clue that previous investigators had totally overlooked. Tucked into the bottom of a long, narrow pencil pocket was a tiny wad of paper. It had obviously undergone numerous washings with the overalls and was blurred beyond all legibility, but by treating it with iodine, Heinrich was able to identify it as a registered-mail receipt numbered 236-L.

The post office traced the receipt for fifty dollars to Roy D'Autremont of Eugene, Oregon, who together with his twin brother, Ray, and another brother, Hugh, had not been seen since October 11, the date of the train holdup. Inquiries about Roy revealed that he was left-handed, had often worked as a lumberjack, rolled his own cigarettes, and was very mindful of his appearance.

Heinrich, meanwhile, had been examining the Colt pistol found near the detonator. Although its exterior serial number had been filed off, he knew that in recent years Colt had taken to duplicating the number inside the gun. Sure enough, when Heinrich dismantled the weapon he found the second engraving. This steered police to a Seattle store where it had been bought by a man calling himself William Elliot. Heinrich opined that, under the microscope, "Elliot's" handwriting and a known example of Roy D'Autremont's were indistinguishable.

All at once, the D'Autremont Gang became the most wanted men in America. Yet despite a fifteen-thousand-dollar reward and circulars that were sent to almost every police department around the world, their whereabouts remained a mystery. And then, in March 1927, an army sergeant studying a wanted poster was struck by the resemblance between Hugh D'Autremont and a soldier he had known in the Philippines. He passed his suspicions on to the authorities. That month, Hugh was captured in Manila and returned to America. In April, his brothers were found working in a steel mill in Steubenville, Ohio, under the alias of Goodwin. All three confessed and were given life imprisonment.

Conclusion

Heinrich's extraordinary work in this case earned him national attention and the nickname "the Wizard of Berkeley." More than anyone else, he made Americans aware of the new face of crime fighting.

Bruno Hauptmann

DATE: 1932
LOCATION: Hopewell, New Jersey
SIGNIFICANCE: Only innovative scientific analysis prevented the "Crime of the Century" from languishing in the unsolved file.

Between seven-thirty and ten o'clock on the evening of March 1, 1932, someone scaled a ladder into a New Jersey bedroom and abducted the nineteen-month-old son of arguably the most famous man alive—transatlantic aviator Colonel Charles Lindbergh. It was a heartless crime that stunned not just America but the world. The only clues were an ill-written ransom demand for fifty thousand dollars signed with a curious logo, some indistinct muddy footprints found beneath the window, the homemade ladder, and a chisel discovered some distance away.

From the outset, the investigation was unorthodox, as Lindbergh chose to put his trust in private parties rather than in official channels. Eventually, an intermediary, Dr. John Condon, contacted the note writer through a Bronx newspaper and agreed to pay the fifty thousand dollars. In a dark cemetery, Condon handed over the ransom in traceable bills and gold certificates to a stranger who spoke with a pronounced German accent. The stranger passed Condon a note saying that the child was aboard a boat moored off the New England coastline, but this turned out to be false. For weeks the investigation foundered, until May 12, when the badly decomposed body of Charles Jr. was found in a wood barely two miles from the Lindbergh estate. The autopsy, conducted by Dr. Charles H. Mitchell, indicated that the child had died from a blow to the head, received in all probability on the night of his abduction.

Shortly after the kidnapping, Arthur Koehler, a scientist with the U.S. Forest Products Laboratory, offered his services to the investigation. By 1932, Koehler was recognized as the world's foremost wood expert, the author of fifty-two government booklets and pamphlets on the subject and one book, *The Uses and Properties of Wood.* He was shown a few slivers of wood from the ladder and asked to identify the type of wood used. This Koehler did, noting in his report the presence of golden brown, white, and black wool fibers, which he speculated might be from the kidnapper's clothing. Then, with police interdepartmental friction mounting to a level that threatened to jeopardize the entire investigation, Koehler found himself sidelined for almost a year.

Eventually, he was summoned to Trenton to examine the ladder in person. His first impression was that it had been poorly constructed; instead of rungs, the ladder had cleats, sloppily mortised with a dull chisel, while other tools, including a plane and a saw, had been used needlessly in places. After making test cuts with the crime scene chisel, Koehler wasn't prepared to state categorically that it had been used on the ladder.

Exhaustive Inspection

Back at the Forest Products Laboratory, Koehler dismantled the ladder into its component parts. Each segment was numbered: the cleats one through eleven, the rails twelve to seventeen. Every mark was noted and indexed. Four types of wood were used: North Carolina pine, birch, Douglas fir, and Ponderosa pine. Marks on rails twelve and thirteen were consistent with wood that had been dressed by a mill using a belt-driven planer set to operate at 230 feet per minute. Four nail holes in rail sixteen, unconnected with the construction of the ladder, suggested that the wood had previously served in another capacity. Because these holes were clean and rust-free, and the wood showed no weathering, Koehler felt sure it had been used indoors, most likely in a barn, garage, or attic.

Although North Carolina pine grew over a very wide region, Koehler thought that because the wood had surfaced in New Jersey, it had probably been milled in the eastern United States. A letter was sent to 1,598 planing mills from Alabama to New York. Of all the mills contacted, only twenty-three reported owning planers configured to the specifications outlined in the letter. Samples of cut wood were requested from each. An example received from the Dorn Lumber Company of McCormick, South Carolina, showed exactly the marks Koehler was looking for.

Because the company had not set its planer to run at 230 feet per minute until 1929, obviously the wood had to have been dressed since then. Company records showed that in that period, forty-six carloads of one-by-four pine had been shipped north of the Potomac.

Arthur Koehler's testimony proved crucial during Hauptmann's trial. (National Archives)

On November 19, 1933—eighteen months after Koehler's initial involvement in the case—detectives finally tracked the wood to the National Lumber and Millwork Company in the Bronx, only to have their hopes dashed when the company admitted that it did not keep records of purchasers.

Bruno Hauptmann was condemned to the electric chair. (National Archives)

This wasn't the first time that the Bronx had figured in the inquiry; earlier, notes from the ransom money had begun to show up in this New York borough. Banks had been given a list of the serial numbers and were told to keep an eye out for them, but not until September 18, 1934—well over two years after the kidnapping—was any significant progress made. A garage attendant who received a ten-dollar gold certificate from a customer jotted down the customer's car license plate number—4U-13-41-NY—on the certificate. When it was deposited at the bank, they immediately contacted the police.

The next day, Bruno Richard Hauptmann, a thirty-four-year-old illegal German alien who worked as a carpenter, was taken into custody. At his Bronx home, police found fourteen thousand dollars of the ransom money. Hauptmann insisted that he was merely minding the money for a friend, Isidor Fisch, who had gone back to Germany and since died from tuberculosis, which explained why Hauptmann was now spending the cash.

Apartment Ransacked

Police tore Hauptmann's home apart. In the garage was a hand plane that produced cuts identical to those found on the ladder. Later, an officer studying the attic, which had been dismantled and shipped to the investigation headquarters, noticed that part of a floor joist had been sawn off. When Koehler laid rail sixteen across this section, the nail holes in the joist and those in the rail aligned perfectly. He later showed, by analyzing both grain and wood, how rail sixteen and the section of board remaining in the attic were originally all one piece.

Samples of Hauptmann's handwriting and spelling were given to Arthur Sherman Osborn, who in 1910 had written a massive textbook titled *Questioned Documents* that became the standard in its field. Osborn and his son, Albert D., also proficient in handwriting analysis, both concluded that Hauptmann had authored the misspelled ransom demands. Oddly enough, right at the outset of the investigation, Scotland Yard, which had become involved in the case because one of Lindbergh's servants was British, had studied the logo and concluded that it was an amalgam of three letters, *BRH,* though not necessarily in that order.

In the overwrought atmosphere that plagued Hauptmann's trial, Koehler's evidence was some of the most lucid and most damaging, as he showed how the plane found in Hauptmann's garage had been used to plane a rung of the ladder. An avalanche of forensic evidence virtually consumed the defendant, resulting in a verdict of guilty on February 13, 1935.

Conclusion

On April 3, 1936, Hauptmann went to the electric chair, still protesting his innocence. Many believe him to be the hapless victim of a frame-up, but while this case has several troubling aspects—most notably the shoddy initial police work—no objective reading of the evidence could fault the jury's conclusion.

John Fiorenza

DATE: 1936

LOCATION: New York, New York

SIGNIFICANCE: In this genuine classic of American forensic science, a hair less than an inch long led investigators to the killer.

Nancy Titterton, thirty-four, lived with her husband, Lewis, at 22 Beekman Place, a Manhattan brownstone popular with the literary set. Each day, Lewis left for his job as an NBC executive, while Nancy worked at home as a book reviewer and a promising novelist. On Good Friday, 1936, two furniture men, returning a couch that had been under repair, climbed the stairs to the Tittertons' fourth-floor apartment. Surprised to find the front door ajar, Theodore Kruger called out. There was no answer. He and his young assistant, John Fiorenza, entered cautiously. They peered into the bedroom. One of the beds showed signs of disorder—a disheveled bedspread—and clothing was strewn carelessly on the floor. Attracted by a light in the bathroom, Kruger slowly opened the door.

Nancy Titterton was lying facedown in an empty bath. Nude except for her rolled-down stockings, she had been strangled with her own pajama jacket. Kruger sent his assistant downstairs to the janitor's office, where he phoned the police.

Such a high-profile victim inevitably prompted a high-powered investigation, and the sixty-five-strong team under the leadership of Assistant Chief Inspector John Lyons got its first break early. When the body was lifted from the bath, a cleanly severed thirteen-inch length of cord came into view. Judging from bruises on Nancy Titterton's wrists, she had been bound before being raped; afterward, the killer had cut the bonds and apparently taken them away with him. But in his haste, he had missed this short piece of cord beneath the body. Lyons ordered his men to check every rope and cord manufacturer in the New York area, little realizing that the search would eventually cover three states and dozens of companies.

The discovery of a smear of fresh green paint on the bedspread—presumably left by the killer—led to vigorous questioning of some workers who had been hired to repaint 22 Beekman Place. Only one, though, had been at work on Good Friday, and his presence elsewhere in the building at the time of the murder—around 11 A.M. according to the medical examiner—was attested to by other tenants.

Lyons was puzzled. Everything pointed to this being a deliberately planned murder—the killer had brought the cord with him and taken it away again—and yet no one in the building had heard anything suspicious. Also, Nancy Titterton was a timid woman, most unlikely to admit a stranger to the apartment, so either the killer had broken in and taken her by surprise, or she knew him. Lyons was inclined to believe the latter.

Unusual Discovery

Dr. Alexander O. Gettler of the city's toxicological laboratory examined the bedclothes. A chemist by discipline, Gettler was well rounded in all aspects of criminalistics, and using a magnifying glass, he studied every stitch of the bedspread. After a while, he found a strand of stiff white hair, barely half an inch long. A more powerful microscope revealed it to be horsehair, of the type used for stuffing furniture. Curious, Lyons obtained a sample of horsehair from the couch that had been under repair. Gettler declared them identical. Because nothing else in the apartment was remotely similar, Lyons concluded that the hair had to be from the sofa; furthermore, because it was too heavy to have been blown into the bedroom by a gust of air, it could only have been transported on someone's clothing. While not entirely ruling out his own detectives as unwitting carriers, Lyons thought it more likely that the horsehair had come from one of the two furniture men. Yet neither had entered the bedroom. Or so they said. But what if one of them had visited the apartment earlier that day?

Lyons visited Kruger's upholstery shop. He found the proprietor there alone. When asked to account for his whereabouts on Good Friday morning, Kruger replied that he had been working in the shop. And Fiorenza? Kruger shook his head slowly. No, his assistant had not arrived for work until 11:50 A.M. He had been delayed for some reason.

Lyons pushed for more. Hesitantly, Kruger admitted that the diminutive twenty-four-year-old Fiorenza had a checkered past but was making every effort to reform. Kruger's next revelation made Lyons even more thoughtful. On April 9, the day before the murder, Fiorenza had accompanied Kruger when he had gone to collect the couch from the Titterton apartment, and this was the second time they had been there. So if Fiorenza had called at the apartment on Good Friday, Lyons speculated, Nancy Titterton would have admitted him as a familiar face.

Back at headquarters, he sent for Fiorenza's record. It showed four arrests for theft and a two-year jail sentence. Of greater significance, though, was a 1934 psychiatric report, in which Fiorenza was diagnosed as delusional and prone to wild fantasies. Such a person, Lyons decided, might well be his own worst enemy.

So began a relentless battle of nerves. Each day, detectives would drop by the furniture shop and chat with Fiorenza about baseball and other topics of the moment, only rarely mentioning the murder. Whenever they did, it was only to say how stymied they were. Fiorenza, encouraged to offer theories about how the murder had occurred, found the attention flattering and let his fertile imagination run wild. How about a lover? Or maybe a pots and pans salesman? The suggestions kept flowing. So, too, did the lapses, as he unconsciously let slip details of the crime that only the killer could have known. Certain in his own mind that Fiorenza was the murderer, Lyons cast around for the proof that would convince a jury.

Breakthrough

He found the evidence he was looking for on April 17: The piece of twine found beneath Nancy Titterton's body had been manufactured by the Hanover Cordage Company of York, Pennsylvania. Their records showed that the distinctive istle-based cord had been widely distributed in the New York area, but eventually detectives traced a wholesaler who had sold a roll to Theodore Kruger's upholstery shop.

On April 21, Lyons ordered Fiorenza to be brought in for questioning. For several hours he resolutely denied all knowledge of the crime; then Lyons produced the length of twine. Fiorenza buckled when he heard how it had been traced to Kruger's store. All of the bluster hissed from Fiorenza's frame. Now the confession came in a rush: how on Good Friday morning, infatuated with Nancy Titterton's beauty and believing her to be equally smitten by him, he had gone to her apartment and gained admittance on the pretext that he was returning the couch.

Once inside, he had inveigled her into the bedroom by suggesting it as an alternative place for the couch; then he had attempted to seduce her. But she had rebuffed his pathetic advances. In a fury, he had overpowered the slightly built woman, rammed a handkerchief into her mouth, tied her hands with cord, and raped her. Afterward, he had knotted the pajama jacket around her throat. He dragged her into the bathroom, cut the cord from her wrists, then dumped her in the bath. He claimed she was still breathing when he left.

Fiorenza's plea of temporary insanity hardly bore close examination. He had obviously considered the likelihood that Nancy Titterton would reject him, otherwise why take along the cord and knife? The prosecution said he had entered the apartment fully prepared to rape, and he had left there a cold-blooded killer. Found guilty, Fiorenza was executed on January 22, 1937.

Conclusion

Gettler's contribution to this case, and that of his assistant, Dr. Harry Schwartz, received extensive press coverage, demonstrating to readers how the miracles of twentieth-century technology were now being used in the business of catching killers.

Samuel Morgan

DATE: 1940

LOCATION: Seaforth, England

SIGNIFICANCE: A single scrap of fabric and the secrets it held provided a landmark case for the advancement of fiber analysis.

At half past six on the night of November 2, 1940, fifteen-year-old Mary Hagan left her home in Waterloo, just north of Liverpool, to buy an evening paper. She did not return. Five hours later, searchers combing nearby railroad tracks found her body in a cement blockhouse. She had been raped and strangled. Beside her lay that evening's edition of the *Liverpool Echo*. Dr. James Firth, head of the Home Office's Northwest Forensic Science Laboratory at Preston, Lancashire, was called out in the middle of the night to examine evidence found at the crime scene. He concentrated his attention on a scrap of muddy fabric found near the body. It looked like a bloodstained finger bandage.

Firth studied the victim's body. On the left side of the neck was what appeared to be an unidentifiable bloody thumbprint. Because none of the victim's injuries had bled, he reasoned that both the print and the bandage belonged to the killer. Most likely the injured assailant had lost the bandage during the struggle and then had pressed his bleeding thumb against Mary Hagan's throat.

Laboratory analysis revealed that the bandage was in two parts: a water-repellent outer layer and an inner bandage impregnated with disinfectant and zinc ointment. The disinfectant—acriflavine—proved to be a valuable clue. Because wartime priorities had restricted its use almost exclusively to military field dressings, Firth suspected that the killer was a serviceman nursing a badly injured thumb. Inevitably, the investigation turned toward the nearby Royal Seaforth barracks, and in particular to a private named Samuel Morgan who had deserted in late September. He was already under suspicion for an earlier attack on another local woman, Anne McVitte.

On November 13, Morgan was detained in London, still showing obvious signs of a badly gashed right thumb, and was returned to Liverpool. At first he was charged only with the attack on Anne McVitte. A prison doctor who examined Morgan's thumb injury thought that it might have been caused by barbed wire and was between one and two weeks old.

When interviewed by the police, Morgan's sister-in-law admitted harboring him since his desertion from the army. She also recalled that on October 31 she had dressed a wound to his right thumb, caused, he said, by some barbed wire. The next day she had applied a fresh bandage, this time adding zinc ointment. On each occasion she had used dressings from Morgan's military kit. Fortunately, she still had some of the bandage and ointment left, both of which she handed over. She denied seeing Morgan since four o'clock on the afternoon of November 2, when he had left her house ostensibly to borrow money from a brother-in-law, James Shaw.

Breathless Meeting

Shaw agreed that he had met Morgan at the Royal Hotel in Seaforth on November 2 between seven and eight o'clock. Yet another brother, Francis Morgan, was also present. At about 7:35 P.M., Samuel Morgan had arrived, panting hard, as if he had been running, and with blood pouring from his thumb. He got Francis to bind the wound with a field dressing, then begged the two men for money as he needed to flee the area immediately. They handed over every penny they had, and he left.

The police asked themselves a question: If Mary Hagan had been killed at around seven o'clock, as they thought, did Morgan have sufficient time to reach the Royal Hotel just half an hour later? A few simple calculations gave them an affirmative answer.

Suddenly, out of the blue, Morgan confessed. He had not intended to kill Mary Hagan, he claimed, just to rob her, but in trying to stifle her screams, he had inadvertently choked her. Singularly lacking in Morgan's statement was any mention of the vicious sexual attack. Given the ephemeral nature of confessions—easily made and just as easily recanted—Firth was asked to intensify his forensic examination.

He compared the bandage found in the blockhouse with the ones obtained from Morgan's sister-in-law. They matched exactly, as did the unopened packet of field dressings that Firth recovered from Morgan's pocket. All three were the same as the dozens of sample dressings supplied by Seaforth barracks, which highlighted a major drawback: In wartime Liverpool, such dressings numbered in the hundreds of thousands; it was hardly proof of murder. Of potentially greater importance was the discovery that soil stains on Morgan's clothing revealed the same trace elements—manganese, copper, and lead—found on the blockhouse floor.

Suddenly, Firth made a critical discovery. As noted, he had already matched the blockhouse bandage to those obtained from the sister-in-law. He now realized that they differed from the other sample bandages in one significant detail. Both had been double stitched, whereas all the barracks' bandages showed a single row of stitching. Firth turned to textile engineer Ronald Crabtree for clarification. Crabtree was in no doubt that the bandage Morgan had worn, the one applied by his sister-in-law, had been handsewn, and none too carefully; every other bandage from the Seaforth barracks was machine-sewn and flawless. Morgan had chosen the only bandage that could betray him.

At his trial for murder, which began on February 10, 1941, Morgan, as anticipated, recanted his confession, claiming that it had resulted from police coercion. But he had no answer for the forensic evidence, and after a weeklong trial, he was condemned to death. He was hanged on April 4, 1941.

Conclusion

Unquestionably, luck played a hand in securing the conviction of Samuel Morgan, but it was the kind of fortune that was becoming ever more common as advances in laboratory analysis continued to seal off the criminal's avenues of escape.

John Vollman

DATE: 1958

LOCATION: Edmundston, Canada

SIGNIFICANCE: This was the first murder case in which neutron activation analysis—so-called atomic evidence—played a decisive role.

Late one May afternoon in 1958, sixteen-year-old Gaetane Bouchard left home to go shopping in the small New Brunswick town of Edmundston. When she had not returned home by eight o'clock, her father, Wilfrid, began telephoning Gaetane's friends, asking where she might be. A few mentioned the name of John Vollman, a twenty-year-old part-time saxophonist who lived just across the border in Madawaska, Maine. Since meeting some months earlier at a local dance, he and Gaetane had enjoyed a casual relationship.

That evening, Mr. Bouchard drove across the border to confront Vollman at the newspaper printing plant where he worked the night shift. The young man seemed guileless enough; he admitted dating Gaetane at one time but denied having seen her recently, not since becoming engaged to another girl.

When Bouchard returned home, there was still no sign of Gaetane. At 11 P.M., he called the police and then resumed his search. On the advice of friends, he checked a disused gravel pit just outside town. Normally a popular spot with couples in cars, tonight it lay deserted. Bouchard parked and began exploring the dark ground, flashlight in hand. Within minutes, the probing beam picked out a suede shoe, instantly recognizable as Gaetane's. Moments later, it fell upon the lifeless body of his daughter. She had been stabbed repeatedly in the chest and back, then dragged off to die in the darkness. Some distance away, a dark pool of blood and tire prints signposted where the murderous assault had begun. While making plaster casts of these prints, an observant police officer noticed two slivers of green paint, one barely larger than the head of a pin, the other somewhat bigger and heart shaped. They appeared to be from a car, probably chipped off by flying gravel as the wheels were accelerated fiercely.

The next day, attempts were made to retrace Gaetane's movements in the hours leading up to her disappearance. At 4 P.M. she had bought some chocolate from a local restaurant. Shortly after this, she was seen talking to the driver of a light-green Pontiac, possibly a 1952, with

Maine plates. An hour later, two friends saw her inside a green Pontiac, in all likelihood the same green car that another witness saw parked by the gravel pit between five and six o'clock. The medical examiner attributed death to multiple stab wounds. There was no evidence of sexual assault, and from the partly digested chocolate remaining in the stomach, he gave the time of death at no later than 7 P.M.

Sexually Aggressive

Possibly through embarrassment at learning of his daughter's apparent promiscuity, Mr. Bouchard had not mentioned Vollman to the police when reporting Gaetane missing. This omission was rectified by friends of the dead girl. Detectives listened with interest as they elaborated on Vollman's reputation as an aggressive small-town gigolo, ill disposed to sexual rejection. Even more illuminating was the information that he had recently purchased a light-green 1952 Pontiac.

Like Mr. Bouchard, the detectives also caught up with Vollman at work. Before interviewing him, though, they checked the parking lot. Vollman's Pontiac was in good condition except for a heart-shaped blemish just beneath the passenger door where the paint was chipped off. When checked, the particle of paint from the crime scene fitted perfectly, a match later verified microscopically.

Vollman treated his inquisitors with shifty indifference, still insistent that he had not seen Gaetane for months. He sneered at eyewitness accounts that placed her in a green Pontiac, pointing out that perhaps the Edmundston police chief ought to be interrogated as well, because he drove an identical vehicle! Even mention of the heart-shaped splinter of paint failed to ruffle his equilibrium.

The Pontiac continued to reveal its secrets. Inside the glove compartment was a half-eaten bar of lipstick-stained chocolate, just like the type bought by Gaetane a few hours before her death. And if that were not enough to convict Vollman, the Edmundston morgue was now set to administer the knockout blow.

A second autopsy had revealed the most compelling clue yet. Entwined in the dead girl's fingers was a single hair, two and a half inches long, most probably pulled from the killer's head as she fought for her life. In order to establish whether this hair came from their chief suspect, investigators turned to a controversial by-product of the atomic age.

Neutron activation analysis (NAA) is a complicated technique in which the sample is first placed in a capsule, then inserted in a nuclear reactor and bombarded with neutrons in order to make it radioactive.

By measuring the rate at which radioactive atoms disintegrate, analysts can identify the sample's trace elements. Using this process, hairs from Vollman and Gaetane were compared to the single hair. In the proportion of sulfur to phosphorus radiation, Gaetane's hair registered 2.02, whereas the sample from Vollman and the single hair were 1.07 and 1.02 respectively. Clearly the single hair had not come from the murdered girl, and just as obviously it was very close to that of Vollman.

Cocksure Defendant

Over strident defense objections, the court decided to admit this revolutionary testimony when Vollman stood trial for murder in Edmundston on November 4, 1958. Having pleaded not guilty, Vollman seemed confident that the jury would disregard what the press had termed "atomic evidence." But as the witness stand filled with scientists eager to explain the new procedure and how it pointed toward the defendant's guilt, the mood in court began to turn palpably against Vollman, so much so that he abruptly announced that he wanted to change his plea to guilty of manslaughter. He now admitted to killing Gaetane, but not intentionally. At first, he said, she had encouraged his advances but then had changed her mind. His last memory was of a struggle breaking out. Everything afterward was a blur, lost in the murk of a blackout.

Vollman's last reckless gamble failed. Found guilty as originally charged and sentenced to hang, he was later reprieved.

Conclusion

NAA is so sensitive that it can identify fourteen different elements in a single inch of human hair. Scientists put the likelihood of two individuals having the same concentration of nine different elements at one in a million. But NAA is costly and of course requires access to a nuclear reactor. With the advent of other identification techniques, it has fallen from favor.

Chester Weger

DATE: 1960

LOCATION: Starved Rock, Illinois

SIGNIFICANCE: Forensic science is only as good as its practitioners and the equipment they use. Here, ineptitude almost set a mass murderer free. Fortunately, other more capable scientists were at hand to avert this near miscarriage of justice.

The savage robbery and murder of three well-to-do Chicago socialites in Starved Rock State Park, Illinois, was a crime that shocked the nation. The women—Mrs. Lillian Oetting, Mrs. Mildred Lindquist, and Mrs. Frances Murphy, wives of prominent businessmen—were found bound with twine and hidden in a cave on March 16, 1960. All three were naked from the waist down. Nearby, on the hard-packed snow, lay a three-foot-long bloodied tree limb, thought to be the murder weapon. Autopsies confirmed that death had resulted from skull fractures and brain damage, but, significantly, there was no evidence of rape, suggesting that the clothing had been deliberately disarranged to throw investigators off the trail.

Because the victims had been staying at the park lodge, detectives concentrated on employees and other guests. For a time the most promising suspect was Chester Weger, a twenty-one-year-old dishwasher at the lodge. Other employees recalled seeing scratches on his face the day the women had been murdered, abrasions that Weger attributed to shaving.

Among his possessions, police found a bloodstained leather jacket, which they sent to their own laboratory for analysis. In the interim, Weger agreed to a polygraph examination, which was carried out by officials from the state police headquarters in Springfield. The results suggested that he was telling the truth when he denied any connection with the brutal crime, and shortly afterward the laboratory report came back—the stains on the jacket were animal blood, as Weger had claimed. Reluctant to give up on Weger as a suspect, the police gave him another polygraph test. Once more, he passed with flying colors.

At this point, Weger was dropped from the investigation, as were all the park employees. Now the police felt that the women had been the victims of a transient thief, especially when a belated inventory of Lillian Oetting's possessions revealed that two rings were missing, presumably stolen by the killer. Local pawnshops were asked to be on the lookout. None reported anything remotely like the missing jewelry. This was hardly surprising, because the elusive rings had not been stolen at all. They were found inside the dead woman's glove, which she had stuffed into her coat pocket before she was killed. Apparently she had feared the worst when approached by a stranger on the trail.

Bungled Investigation

By now it was clear that the investigation had been less than thorough. As disgusted police officers followed the furor over the ring incident, which culminated in the laboratory superintendent's resignation, they pondered the likelihood of other mistakes having been made. This brought about a renewed interest in the twine used to bind the women, originally thought to be of too common a type to be traced. Deputies found that it was a twenty-strand variety and compared it with other twine used at the lodge, which also had twenty strands. The manufacturer, a Kentucky firm, confirmed that the only place in Illinois to buy that particular cord was Starved Rock State Park.

Officers again targeted the lodge. In a toolshed they not only found lengths of the twine but also knots in the twine that were identical to those used to bind the murder victims. A short list of people who used the shed was drawn up. High on that list was Chester Weger. Again his jacket was submitted for analysis, this time to the FBI Laboratory in Washington. Scientists there made no mistake—the stains were human blood, most likely from the same group as one of the victims.

Once again, Weger became the chief suspect. And yet twice he had passed a polygraph. Even so, it was decided to test him again. Top polygraph expert John Reid of Chicago conducted the examination. This time, the machine clearly indicated deception when Weger denied committing the murders.

Other evidence slowly came to light. In the fall of 1959, not long before the triple murders, a girl had been raped and bound with twine exactly like the cord used in the murders. From a photo layout, she identified Weger as her assailant. The girl told police that during the attack Weger had kept a bullet clenched between his teeth, a grim reminder, he warned, of what would happen to her if she ever told anyone.

Confronted with this new evidence, plus the revised blood analysis and the incriminating polygraph findings, Weger confessed. On March 3, 1961, Weger was convicted of murdering Lillian Oetting and sentenced to life imprisonment. It was considered unnecessary to prosecute Weger for the other two crimes because he was already under lock and key.

So how did Weger fool the polygraph? Reid, asked to account for his success where others had failed, expressed it as follows: "Simple enough, it's a matter of technique, in knowing how to conduct such an experiment effectively."

Asked the same question, Weger had an entirely different rationale. "Before the first two tests," he said, "I just swallowed a lot of aspirin and washed it down with a bottle of Coke. That calms a guy down, you know. Why I didn't do that before this other guy tested me I'll never know."

At the time of this writing, Chester Weger is still behind bars at the Menard Correctional Center. In September 2005, the Illinois Prisoner Review Board unanimously rejected his latest request for parole. They are not scheduled to reexamine his application until 2008.

Conclusion

Some good actually came of the Starved Rock Murders. The amateurish Illinois state crime laboratory, then no better equipped than a high school chemistry lab, was abolished by the state legislature. In its place, Illinois today has what ranks among the finest state crime laboratory systems in America.

Stephen Bradley

DATE: 1960

LOCATION: Sydney, Australia

SIGNIFICANCE: To solve a crime that stunned the nation, Australian police put together a forensic team made up of geologists, doctors, scientists, and, most important, the general public.

Until June 1, 1960, Bazil and Freda Thorne were an obscure middle-class couple living in suburban Sydney, Australia. On that day their anonymity was forever shattered by the news that they had won one hundred thousand Australian pounds ($210,000) in the state lottery. Any exhilaration they felt lasted all of six weeks: On July 7, someone snatched their eight-year-old son, Graeme, from the street as he made his way to school.

Later that morning, a man with a heavy foreign accent phoned the

Thorne household and demanded twenty-five thousand pounds for the boy's safe return. He called again at 9:40 P.M., repeated his demand, and hung up quickly, thereby foiling police attempts to trace the call. That would be his last communication.

The next day, Graeme's empty school case was found in a garbage dump in a distant part of town. On July 11, his cap, raincoat, arithmetic book, and lunch bag turned up another mile away. But there was no sign of the boy. An eyewitness report of a 1955 blue Ford Customline parked outside the Thorne residence on the morning of the kidnapping offered hopes for a lead, but tempering official optimism was the fact that five thousand such vehicles were registered in the state of New South Wales.

Each passing day diminished hopes of finding Graeme alive. And so it proved. Five weeks later, on August 16, his body was found wrapped in a rug about ten miles from his home. He had been suffocated and clubbed to death. The Scientific Investigation Bureau was responsible for extracting the trace evidence. Certain items of Graeme's clothing had been encrusted with a pink soil-like substance; its significance would become apparent only later. The rug, which had one tassel missing, was covered with both animal and human hairs, as were the victim's jacket and trousers. Around the shoes and socks, a mold culture had begun to form. Further traces of botanical matter were found in the clothing.

All these items were subjected to a barrage of tests. Dr. Cameron Cramp of the Government Medical Office identified four different kinds of hair: three human and one, he said, almost certainly from a Pekingese dog. The shoe mold, said Professor Neville White, comprised four types of fungi. By replicating as closely as possible the circumstances of Graeme's death, he dated the mold as six weeks old, suggesting that the boy had been killed immediately after abduction. The pink substance was identified by Horace Whiteworth, curator of the Geological and Mining Museum in Sydney, as a kind of mortar frequently used in Australian house construction.

All of this left only the botanical material. Because of the variety and amount, the entire scientific staff at the National Herbarium was called upon to lend assistance. By mid-September, all of the various leaves, seeds, and particles of twigs found on the body had been identified. One, an extremely rare cypress seed, offered the most hope of providing a lead, because it definitely did not grow in the vicinity where the body was found.

Public Appeal

Investigators now reevaluated the pink mortar; if they could locate a housing development in which it coexisted with the rare cypress, a valuable link might be established. An appeal broadcast to the general public, urging anyone who knew of such a property to contact the police, brought immediate results. A mailman suggested that they try a particular house on his route. The pink house at 28 Moore Street, Clontarf, certainly fit the bill: on either side of the garage door grew not only the rare cypress but also other plants whose traces had been found on Graeme's body. The current residents, having moved into the house just recently, were soon eliminated as suspects, but not so the previous tenant.

Stephen Bradley (whose real name was Istvan Baranyay), a Hungarian immigrant, had moved out of the house on the very day of the kidnapping. Neighbors confirmed that he spoke heavily accented English, owned a blue Ford Customline, and was devoted to his Pekingese dog. Inside the house, detectives found an old photograph. It showed Bradley and his family enjoying a picnic, sitting on the very rug in which Graeme's body had been wrapped. They even found the rug's missing tassel, torn off and left unnoticed on the floor.

News that on September 20 Bradley had sold the Ford to a used-car dealer sent officers rushing to the dealer in question. There on the lot stood the car they had been seeking for two months. Scientists examined every inch. In the trunk were pink particles identical to those found on Graeme's body.

Through travel agents, police learned that Bradley; his wife, Magda; and their three children had booked passage for England aboard the liner *Himalaya*. The ship had sailed on September 26 and was currently en route to Colombo, Ceylon. Before leaving, he had left his dog at a veterinary hospital for onward transmission to England. When Dr. Cramp examined hair from the Pekingese, he declared it to be indistinguishable from the hair found on the dead boy's clothes.

Detectives flew to Colombo, and were waiting at the quayside when Bradley's ship docked. After a monthlong legal wrangle—Ceylon had no formal extradition treaty with Australia—Bradley was returned to stand trial. During the flight, he claimed that Graeme had suffocated accidentally in the trunk of his car, but he failed to account for how the boy's skull had been crushed. Later, Bradley recanted this confession, to no avail. On March 29, 1961, he was sentenced to life imprisonment. Four years later, his wife divorced him and went to live in Europe. Bradley's life sentence ended on October 6, 1968, when he dropped dead from a heart attack at forty-three.

Conclusion

Because Australia was unused to kidnapping—the only other instance had occurred thirty years previously—Graeme Thorne's plight started a media feeding frenzy. Within an hour of his abduction, every news station in the country had run the story. Whether this prompted Bradley to panic and kill the boy is unknowable; more certain is the likelihood that such hysteria in no way aided the missing lad or those charged with finding him.

Roger Payne

DATE: 1968

LOCATION: Bromley, England

SIGNIFICANCE: As a fiber-gathering exercise, the forensic work carried out in the aftermath of this horrific murder is virtually unparalleled.

At a few minutes past eight o'clock on the night of February 7, 1968, newly married Bernard Josephs arrived home at his flat in Bromley, Kent, to find the place in darkness. He let himself in, switched on the lights, and put his raincoat and briefcase on the settee, all the while calling out for his wife, Claire. Puzzled by the strange silence, he entered the bedroom. He didn't see his wife at first. Then he saw legs protruding from under the bed. Claire was lying facedown on the floor, her throat slashed wide open. She was wearing, among other clothing, a cerise woolen dress. Although no one realized it at the time, this dress—a present from her husband—would play a critical role in trapping Claire's killer.

The autopsy, performed by Professor James Cameron, confirmed the frenzy of the knife attack. Her throat had been severed to the spine. Defensive wounds on the right hand showed how hard Claire had fought for her life. It appeared as though the killer had half-strangled her first, then grabbed a knife with a serrated blade. Cameron put the time of death at approximately 7:15 P.M. There was no evidence of a sexual assault. No weapon was found at the flat, but the next morning a bread knife with an eight-inch serrated blade was noted to be missing.

The inquiry, headed by Detective Superintendent John Cummings, began piecing together the last few hours of Claire's life. At 5:45 P.M., her mother-in-law had phoned with a recipe for lemon soufflé. The two women had talked for about fifteen minutes, and sometime between 6 and 7 P.M. Claire had begun preparing the dish. Because the ingredients were still in a mixing bowl in the kitchen, Cummings reasoned that Claire had been interrupted by her killer. He further thought that given the absence of any evidence to indicate forcible entry, this interruption had taken the form of a visitor whom she knew.

This was borne out by the half-empty coffee cup in the kitchen and a plate of cookies. Cummings, noting that the cookies had been taken from a freshly opened packet that lay on the sideboard, took this to mean that the caller was unlikely to be a close friend. Such formality was not Claire Josephs's way with people she knew well.

Violent Background

By concentrating on friends and relatives, Cummings soon came up with the name of Roger Payne, a twenty-six-year-old bank clerk who had previous convictions for attacking women. Payne and his wife were acquaintances of the dead woman and had visited her apartment once in January 1968. What Claire wore during this visit—a brown skirt and a green sweater—would later assume great significance. One week later, January 15, she began work as a telephone operator, and only then did she start wearing the cerise woolen dress in which she was found the day she died. Prior to this, it had been stored in a suitcase and had been there since mid-1967.

When interviewed, Payne had numerous scratches on his hands, the result, he said, of a quarrel with his wife a few days earlier. He was asked to account for his movements on February 7. That morning, wearing a freshly pressed suit from the cleaners, he had gone to London for a job interview. On his way home, he decided to visit his mother, only for his car to break down on the way. While he was tinkering with the engine, the hood slipped and fell, which explained the bruise on his forehead that Cummings had noticed. After some time, he got the car running and carried on toward his mother's. According to her first statement, Payne arrived sometime after six o'clock and stayed only a few minutes; Payne's version had him showing up at 6:30 P.M. and leaving an hour later. The final leg of his journey was again marred by car trouble. This time it proved more difficult to fix, and he did not arrive home until 9:15 P.M. The next day, he returned his suit to the cleaners.

Cummings didn't believe a word of it and asked Margaret Pereira, senior principal scientific officer at the Metropolitan Police Forensic Science Laboratory, to analyze Payne's clothing. His overcoat and hat both showed traces of blood. Pereira typed the blood as group A, the same as Payne's (Claire Josephs belonged to the much rarer blood group AB). The coffee cup found at the crime scene was more enlightening. Saliva samples taken from around the rim showed traces of blood group A. While this meant that Payne could have drunk from the cup, so might have 42 percent of the general population.

Any ambiguities raised by the serological findings were more than outweighed by the sheer volume of fiber evidence that was uncovered. Claire's distinctive cerise dress was made from a fluffy woolen fabric, the kind that sheds easily. These fibers—fluorescent and easily detected under ultraviolet light—made a mockery of Payne's precautionary trip to the dry cleaners. The surfaces were clean, but not the seams and hems. Sixty-one cerise fibers were recovered from his clothing, and he could not claim to have picked up the fibers from his visit to the Josephs' apartment in January. As noted, Claire did not start wearing this particular dress until a week after the Paynes called.

More Damning Evidence

Such success in one direction prompted Pereira to explore the possible transference of fibers from Payne's clothing to that of the victim. The cerise dress proved disappointing; its woolly texture made it less than ideal for the examination of contact fibers. But on Claire's raincoat, hung up just inside the front door, were a number of fibers that might have come from Payne's overcoat and others that matched his suit. Because all of the fibers were of a fairly common type, it was impossible to be more specific. At Payne's house, detectives found a frayed red scarf. Twenty rayon fibers matching this scarf were found on Claire's raincoat. Another fiber of the same type was recovered from her left thumbnail.

With such a wealth of fiber evidence, it was time to extend the search to Payne's car. Cummings reasoned that when Payne drove his Morris 1100 away from the apartment, he took the knife with him for later disposal. If that was the case, then the likeliest hiding place for the knife would be inside the pocket on the driver's door. Sure enough, tests at the bottom of the pocket found traces of group AB blood. This was further isolated by testing for the presence of PGM (phosphoglucomutase) groups. The resulting combination of AB, PGM 2-1—the same as Claire Josephs—occurs in just one person in eighty. Clearly Payne had tried to wash the inner surface of the driver's door, but very faint human bloodstains were detected in various places in the washed area. Debris on the car floor revealed several fibers matching the nylon carpet in the victim's apartment, together with yet more fibers from her dress.

Few defendants have faced such a formidable forensic barrage as Payne did when he stood charged with the murder of Claire Josephs. It was an unanswerable case, one his attorneys found impossible to overcome. Convicted on May 24, 1968, he was sentenced to life imprisonment.

Conclusion

The investigation of this murder has become a textbook classic of forensic science. This was to be Detective Superintendent Cummings's last major case. He always considered it his most successful.

Jeffrey MacDonald

DATE: 1970

LOCATION: Fayetteville, North Carolina

SIGNIFICANCE: Often overlooked in the notorious *Fatal Vision* murders is how an FBI analyst overcame crime scene deficiencies to provide the prosecution with its deadliest line of attack.

An emergency phone call received at 3:40 A.M. on February 17, 1970, sent military police racing to the Fort Bragg home of army doctor Captain Jeffrey MacDonald. The scene that greeted them was horrific. MacDonald and his wife Colette lay sprawled on the master bedroom floor. Colette was obviously dead, a torn blue pajama jacket full of holes draped across her chest. Beside her, clad only in blue pajama bottoms, MacDonald lay motionless but conscious. Above them, daubed in blood on the bed headboard, was the word PIG. Down the hallway, their two young daughters had been hacked to shreds.

In the midst of this carnage, with military police (MPs) rushing headlong about the house and inadvertently trampling evidence underfoot, MacDonald was an icon of lucidity. He told of falling asleep on the living room couch, only to be awakened by Colette's frantic cry. In the darkness he vaguely made out four hippies standing over him, three men and a woman who kept chanting, "Acid is groovy . . . kill the pigs." MacDonald hurled himself at the intruders. They responded by slashing him with an ice pick. To ward off the blows, MacDonald wrapped his blue pajama jacket around his hands; even so, he sustained multiple stab wounds and was quickly knocked unconscious.

Sometime later, he came to. Stumbling from room to room, he attempted mouth-to-mouth resuscitation on each of his daughters, but to no avail. Then he found Colette. He pulled a small knife from her chest and tossed it onto the floor. Next he draped his pajama jacket over her. Then he phoned for help.

It had been a frenzied attack. Colette had been stabbed more than twenty times in the chest, her head pulverized by half a dozen blows, both arms broken in a vain attempt to fend off the blitz. Two-year-old Kristen received thirty-three stab wounds, while Kimberly, five, succumbed to a rainstorm of blows and knife wounds. By contrast, MacDonald's injuries were relatively minor, and before being removed for medical treatment, he was able to furnish detailed descriptions of his attackers.

Afterward, Detectives William Ivory and Franz Grebner studied the room where a life-and-death struggle had allegedly taken place. Something wasn't right. Next to the couch where MacDonald said he had been attacked, a lamp remained upright, and apart from a coffee table that was turned over on top of some magazines and an empty flowerpot in the middle of the floor, the disorder was minimal. To the experienced investigative eye, it had a distinctly staged appearance. Also, why had the gang allowed MacDonald—the only eyewitness—to live?

Manson Copyists

Perhaps, mused Ivory, because there was no drug-crazed gang? Had it all been an invention of MacDonald's? There was ample reason to think this. Among the magazines on the floor was the latest edition of *Esquire,* which featured extensive coverage of the recent Manson massacres in southern California. The similarities between that slaughter and this—homicidal dropouts high on hallucinogenic drugs—were starkly obvious. If MacDonald had been seeking to mask his own murderous activities, then such an article might well have provided the inspiration.

Another troubling aspect was MacDonald's poor eyesight; he

needed glasses to read and drive a car. Supposedly he had been fast asleep in the pitch-black living room when he heard his wife cry out. How, in those few tumultuous seconds before being knocked unconscious, was he able to glean such good descriptions of his attackers? Moreover, the house was still in darkness when the MPs arrived. Wouldn't someone dialing for help on the telephone at least want the light on?

There were other, more tangible reasons for doubting MacDonald's story. Among the bloodstained bedding was the finger section of a latex glove, the kind that surgeons wear. It had been torn, as though removed in a hurry, possibly by the person who wrote PIG on the headboard. Also, the knife MacDonald claimed to have pulled from his wife's chest and then tossed onto the carpet was free of fingerprints. But the most damaging evidence came to light when Colette was lifted from the bedroom floor.

Beneath her body were several blue threads. Yet MacDonald claimed that he had laid his blue pajama jacket on top of Colette after finding her. Eighty-one blue fibers were recovered from the master bedroom. Still more were peeled off a sizable chunk of bloodstained wood found outside the back door. Beneath a nearby bush, detectives found a knife and an ice pick, both wiped clean.

Those ubiquitous blue threads kept turning up—nineteen more in Kimberly's bedroom, including one beneath her fingernail, and two in Kristen's bedroom. Microscopic analysis matched all of them to MacDonald's pajama jacket. Curiously, the only room where investigators failed to find any such threads was the living room, where MacDonald claimed to have fought for his life.

What they did find was blood, pints of it, splashed around all three bedrooms. Against all the laws of probability, each family member had a different blood group, making it possible to track the movement of the victims around the house, especially Jeffrey MacDonald. The only two places where his blood occurred in any significant amounts were the kitchen—by a cabinet containing a supply of surgical rubber gloves—and the bathroom sink, where, investigators speculated, MacDonald had stabbed himself after butchering his family.

Just as enlightening were the bloodstains the investigators didn't find. According to MacDonald, he had used two telephones to call for help; neither showed any sign of blood or even fingerprints. And the place where he fell in the hallway after being stabbed was also free from blood. In the living room, the only traces of MacDonald's blood were on a pair of spectacles and the copy of *Esquire* magazine. Not much for a man who claimed to have been attacked with an ice pick.

On May 1, the army felt it had enough to charge MacDonald with triple murder, but subsequent inquiries revealed such a catalog of official incompetence—many samples of vital trace evidence had been misplaced or lost completely—that in October 1970, all charges were abruptly dropped.

After leaving the army, MacDonald appeared on a TV talk show; joking almost, he berated the military for their attitude toward him. At the same time, he relegated the fate of his family to a few disinterested sentences. It was an oddly callous performance, and more than anything else, this was responsible for the renewal of official interest in Jeffrey MacDonald.

FBI Brought In

His pajama jacket was sent to the FBI Laboratory in Washington, D.C., where analyst Paul Stombaugh made an interesting discovery: all forty-eight holes made by the ice pick were smooth and cylindrical. In order for this to have happened, the jacket would have had to remain stationary. How, he reasoned, if MacDonald had this jacket wrapped around his hands to defend himself and was dodging a torrent of blows, could such a thing have occurred? Equally incongruous was a large stain made by Colette's blood that covered two parts of the torn jacket. When the parts were held together, the stain matched up exactly, suggesting that the stain had been made before the jacket was torn, a fact inconsistent with MacDonald's statement that he had placed the jacket over his wife only after finding her dead.

Also, by folding the jacket in one particular way, Stombaugh showed how all forty-eight tears could have been made by twenty-one thrusts of the ice pick, coincidentally the same number of wounds that Colette had suffered. This evidence, presented to a grand jury in July 1974, resulted in three murder indictments against Captain Jeffrey MacDonald.

Delays, motions, and countermotions kept the case in limbo for years, but on July 16, 1979, nine and a half years after the murders, MacDonald finally faced his accusers in a North Carolina courtroom.

Unquestionably, the dramatic highlight came when prosecutors Brian Murtagh and James Blackburn abruptly staged an impromptu reenactment of the alleged attack on MacDonald. Murtagh wrapped a pajama top around his hands and tried to fend off a series of ice-pick blows from Blackburn. For his troubles, Murtagh received a small wound to the arm, but two telling points had been made. First, all of the holes in the pajama top were rough and jagged, not smoothly cylindrical as the holes in MacDonald's pajama jacket had been. Second, Murtagh was stabbed, albeit not seriously. When MacDonald had arrived at Womack Hospital, he did not have a single wound on his arms.

In his own defense, MacDonald offered little except petulant arrogance; today he is serving three consecutive life terms.

Conclusion

MacDonald has no shortage of supporters who believe him to be innocent. Other observers highlight the forensic evidence—inexpertly gathered, maybe, but in its totality convincing proof that justice, although delayed, was served.

Lionel Williams

DATE: 1976

LOCATION: Hollywood, California

SIGNIFICANCE: In this widely publicized murder, the preservation of physical evidence helped overcome a profusion of mistaken eyewitness testimony.

In his teens Sal Mineo had been one of Hollywood's hottest actors, but like many precocious stars before him, he found graduation to adult roles difficult, and his once-flourishing career went into a steady decline. As screen parts grew scarcer, he turned to the theater. On the night of February 12, 1976, Mineo was on his way home from directing a stage play. He had just parked his blue Chevelle behind the Hollywood apartment block where he lived when neighbors heard a cry of "Oh God, no! Help! Someone help!" followed by sounds of a struggle, more screams, then silence.

Seconds later, bystanders saw a man flee down the driveway and speed off in a car—a white man with long brown hair, they later told police. By the time they reached the thirty-seven-year-old Mineo, he was near death, a long stream of blood flowing from a stab wound in his chest. All efforts at resuscitation failed, and within minutes Mineo expired.

At first it was thought that Mineo's recent charitable work with prison reform had backfired badly and that an ex-con had killed him. That hypothesis crumbled the moment officers entered Mineo's apartment. The walls were covered with photos of nude men, and stacks of homosexual literature filled the bedroom. The investigation turned around completely. Now a lover's quarrel seemed the likeliest motive. Police scrutinized the photos, searching for someone whose description matched that of the brown-haired white man seen running from the apartment block shortly after Mineo's cries for help were heard. They found nothing.

The autopsy was conducted by Dr. Manuel R. Breton, deputy medical examiner of Los Angeles County. After establishing the cause of death, he X-rayed the lower chest and upper abdomen to see if any metallic fragments from the knife could be found. There were none. As a precautionary measure, he ordered that the chest section pierced by the knife be stored for possible later use at the Los Angeles County Forensic Science Center. As it happened, Dr. Breton's circumspection helped trap a killer.

For two years detectives chased every lead, only to see them all disintegrate. And then word reached them of a man imprisoned in Michigan on a bad-check charge, twenty-one-year-old Lionel Williams, who had boasted to a prison guard of killing Mineo. Earlier in Los Angeles, his wife, Theresa, had told police that on the night of the killing her husband had come home drenched in blood and admitted stabbing Mineo with a hunting knife.

Brutal Muggings

To no one's surprise, Williams soon retracted his story, but police in Los Angeles had already unearthed evidence linking him to a chain of vicious muggings in the city. Now they wondered if Mineo's death had resulted from a random mugging. Despite a witness's description of the getaway car as a yellow subcompact (police had discovered a loan agreement showing that Williams was driving a yellow Dodge Colt on the night of the killing), one giant roadblock was holding up the investigation: Every eyewitness had described a brown-haired white male fleeing from the crime scene—Lionel Williams was *black*, with an Afro hairstyle!

Then the police remembered the stored chest section. At the L.A. County Forensic Science Center it is customary, when a stab wound causes a fatal injury, to examine the wound for various characteristics such as length, width, thickness, and whether the knife used was a single- or double-edged blade, sharp or dull. The wound is also dissected surgically, layer by layer. Through this procedure it is possible to create a cast of the wound itself, which provides a precise means of identifying the murder weapon, should it ever be recovered, by matching the wound with the knife. At the time of the Mineo murder, the center had acquired a number of knives to assist their research. With a description from Mrs. Williams of the hunting knife owned by her husband—she even knew its price, $5.28—they were able to find an identical knife in their collection.

Normally, the act of inserting an allegedly matching knife into a wound during an autopsy is frowned upon, because it might distort the incision. But because this particular chest section had been fixed in formalin for storage, it was possible to do so without compromising the specimen. When inserted into the wound, the blade of this knife matched perfectly.

Although the scientific case seemed compelling enough, there was still the hurdle of eyewitness identification to overcome. Now Williams's past came back to haunt him. Tucked away in police files was a mug shot taken some years earlier, when he had been suspected of another crime. This photograph showed Williams, not sporting an Afro, but instead with long processed hair, dyed light brown and worn in a Caucasian style. Instantly the reason for the eyewitness confusion was crystal clear. At his subsequent trial, Williams was found guilty and sentenced to life imprisonment.

Conclusion

Eyewitness testimony can be notoriously inaccurate, a constant problem for the courts. Here it presented a seemingly insoluble obstacle to the prosecution, and had it not been for some excellent scientific work, detectives might still be seeking Sal Mineo's killer.

Wayne Williams

DATE: 1979

LOCATION: Atlanta, Georgia

SIGNIFICANCE: Fiber evidence allied to probability analysis underpinned the prosecution's claim that Wayne Williams was the worst child killer in U.S. history.

By 1981, the city of Atlanta was in the grip of a two-year killing spree that had left more than twenty young black males strangled or otherwise asphyxiated. Just about the only clues were fibers found on the bodies and clothing of the victims; these were forwarded to the Georgia state crime laboratory. Two types were isolated: yellow-green nylon and violet acetate. The yellow-green fibers were very unusual, coarse with a trilobed cross-sectional appearance, of a kind associated with carpets or rugs. Despite intensive efforts, it was not possible to determine the manufacturer.

In February 1981, following newspaper accounts of this fiber analysis, the killer began dumping his victims in rivers. They were also now either nude or nearly so. Not for the first time, a serial killer seemed to be monitoring media accounts of his activities and modifying his modus operandi accordingly.

In the early hours of May 22, 1981, police staking out the Chattahoochee River, one of the killer's favorite dumping grounds for his victims, heard a loud splash. They spotted a station wagon on the James Jackson Parkway Bridge. The driver, Wayne Williams, a chubby twenty-three-year-old black music promoter, was questioned but allowed to leave. Two days later the body of Nathaniel Cater, twenty-seven, was dragged from the river a mile downstream; in his hair was a single yellow-green carpet fiber. Suddenly the police wanted to know a good deal more about the plump entrepreneur. On June 3, they obtained a search warrant for Williams's house and car. Throughout his home, the floor was covered with yellow-green carpeting.

With little actual evidence to connect Williams with the Atlanta slayings—apart from his reputation as an aggressive homosexual—it became apparent that this case would be decided in the laboratory. Working with chemists at DuPont, America's largest producer of fibers, FBI analysts passed the yellow-green fibers recovered from the victims through a spinneret, a device that stretches the sample, giving it distinguishable optical characteristics. The unique trilobed cross-section enabled them to determine that it had been manufactured by Wellman Inc., a Boston textile company.

Company records showed that this particular fiber—called Wellman 181B—had been made and sold to various carpet makers during the years 1967 through 1974. Because each carpet manufacturer has its own dyes and weaving techniques, it was possible to track these fibers to the West Point Pepperell Corporation of Dalton, Georgia. They had manufactured a line of carpeting known as Luxaire, and one of the colors offered was English Olive, which both visually and chemically matched the carpeting found at Williams's home.

Statistical Probabilities

Although this undoubtedly cast even greater suspicion on Williams, it was far from conclusive proof of guilt. After all, what were the odds of the fibers having come from his carpet and his alone? Only the prevalence of Luxaire English Olive could answer that. For instance, if every house in the Atlanta metropolitan area were fitted with Luxaire English Olive, then its evidential value would be nil. On the other hand, if that particular carpet were found in Williams's house and nowhere else, then any reasonable jury would have to regard that as highly significant. Somewhere between these polar extremes lay the statistical probability.

Although Luxaire was manufactured from 1970 through 1975, it used Wellman 181B fiber for only one twelve-month period between 1970 and 1971. During that time, 16,397 square yards of Luxaire English Olive were sold by West Point Pepperell to retail outlets across ten southeastern states, including Georgia. Compared with the 1979 total US residential carpeted floor space, estimated by DuPont at 6.7 billion square yards, this was a minuscule amount of carpet.

These were the indisputable facts. In order to establish the statistical probability of Luxaire English Olive being found in any one residence, certain conservative assumptions had to be made: first, that sales of Luxaire English Olive were evenly distributed throughout all ten states, and that it was installed in only one room, average size twenty square yards. Erring on the side of caution, investigators next assumed that all of the ten-year-old carpet was still in use, unlikely because the average life span of commercial dwelling carpet is approximately five years. This allowed them to calculate that in Georgia one could expect to find eighty-two homes containing the carpet. Because at the time of Williams's arrest there were 638,995 occupied housing units in metropolitan Atlanta alone, the odds of randomly selecting a home in that city with one room carpeted in Luxaire English Olive were 1 in 7,792—a very low chance indeed. Put another way, in order to randomly pick up the fiber found in his hair, Nathaniel Cater would have had to visit almost eight thousand houses in Atlanta or just one—the home of Wayne Williams.

The Killing Car

Although Williams was suspected of as many as twenty-eight murders, prosecutors felt their best chances of conviction rested with just two cases: Nathaniel Cater and Jimmy Ray Payne, whose seminude body had been recovered from the Chattahoochee River on April 27, 1981. Again, they were pinning their hopes on fiber analysis: A single fragment of rayon, found on Payne's shorts, had been matched to the carpeting in Wayne Williams's 1970 Chevrolet station wagon. Once Chevrolet provided details of all the pre-1973 cars that had been fitted with this kind of carpet, the Georgia licensing authorities clarified how many of these vehicles were registered in the Atlanta metropolitan area in 1981. The answer was 680. Of the more than two million cars registered, this represented a ratio of 1 in 3,828, which was also the statistical likelihood of Payne having acquired the fiber by random contact with any other car except that belonging to Williams.

At this point, the odds against Cater having picked up the fiber randomly, 1 in 7,792, were multiplied with the possibility that the rayon fragment found on Payne's shorts had an equally random origin, 1 in 3,828. The laws of probability say that this will occur no more than once in every 29,827,776 times! Wayne Williams was trying to buck some awfully big numbers.

But still the prosecution wasn't satisfied. During the trial they introduced into evidence a chart showing twelve other victims of the Atlanta child killer. On each body were found between three and six fibers that could be traced to either Williams's home or automobile. The odds against such a likelihood—virtually incalculable—ran into the trillions. To help the jury make sense of these mind-boggling figures, the prosecution prepared more than 40 charts and 350 photographs.

It was a job well done. The anticipated lengthy jury deliberation actually took less than twelve hours. On February 27, 1982, Williams was convicted and sentenced to two life terms.

Almost inevitably in such an extraordinary case, the jury's verdict by no means marked the end of the case. Campaigners insisted that Williams had been railroaded by a racist police force determined to pin the crimes on a black man, rather than investigate allegations of a reported Ku Klux Klan involvement. An absence of publicly available evidence to support these claims was also blamed on the state, with Williams's supporters contending that the state had deliberately suppressed evidence that could have led to an acquittal. This argument finally received a judicial airing in 2005. After listening carefully to arguments from both sides, U.S. district judge Beverly Martin delivered her decision on February 8, 2006. She wrote that none of the allegedly withheld evidence "would have had more than a minimal impact upon the outcome of Mr. Williams's trial had it been presented to the jury." Given the inflammatory nature of the Atlanta killings, it is unlikely that this will be the final word.

Conclusion

Although there was eyewitness evidence linking Williams to one of the victims, the use of probability theory—something never before employed with fiber evidence in a trial—was nevertheless crucial in securing a conviction. Those who have complained that Williams was tried more by the law of probabilities than the law of the land would do well to bear in mind that for the most part, juries are composed of sensible citizens, and that there are only so many coincidences they are prepared to swallow.

John Joubert

DATE: 1983

LOCATION: Bellevue, Nebraska

SIGNIFICANCE: A length of rope that originally defied identification ultimately provided irrefutable evidence linking a serial killer to one of his victims.

After making just three stops on his Sunday morning newspaper route, Danny Joe Eberle, a thirteen-year-old Nebraska schoolboy, vanished. Police in his hometown of Bellevue, just south of Omaha, organized a massive search, and on the following Wednesday, September 21, 1983, his badly mutilated body was found in some high grass alongside a gravel road, three miles from where he had been last seen.

He had been stripped and bound with rope. Knife wounds all over the body suggested that he had been tortured before death.

Details of the crime were passed to the FBI's Behavioral Science Unit, which compiled a provisional profile of the likely killer: a white single male, about twenty years old, a loner, a latent homosexual, small and neat in appearance, with a fairly slender build. This last characteristic was suggested by the body's proximity to the road; profilers reasoned that the killer was physically incapable of carrying the body far, and had been forced to dump it in a well-trafficked area.

The rope used to bind Danny Joe seemed to offer the best chance of a lead. Generally, the more unusual the rope, the more useful it is in evidence, because a product available widely is difficult to link to an individual. But here the rope was so unusual that nobody could recognize it. Even the FBI declared themselves baffled as to its origin.

On December 2, another young boy, Christopher Walden, thirteen, the son of an officer at nearby Offutt Air Force base, was abducted. Three days later, two hunters found his body in a heavily wooded area. The hideous nature of his stab wounds made it abundantly clear that the killer had struck again. There were two sets of footprints in

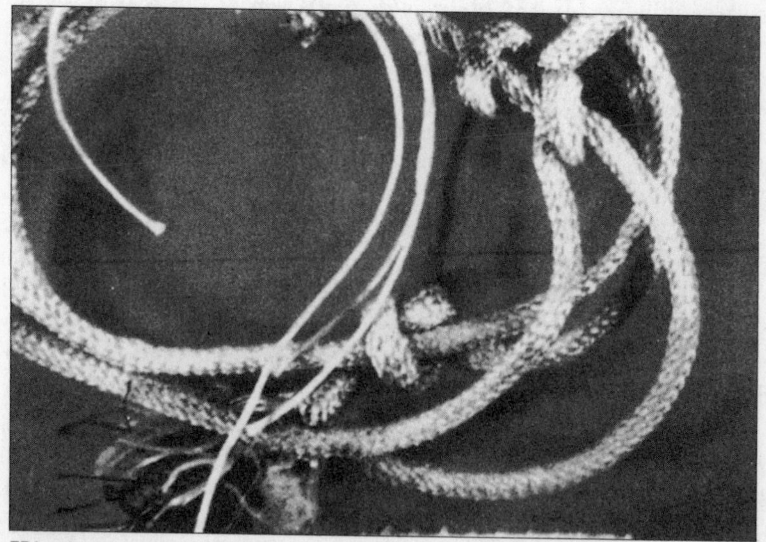

FBI agents tracked these few strands of rope around the world to serial killer John Joubert.

the fresh snow leading to the body and only one set leading away. Such familiarity with the terrain made it likely that the killer was someone local, possibly an airman on the base, and the ease with which both victims had been abducted led profilers to believe that he was someone involved with young boys in some way—perhaps the Boy Scouts, Little League, or maybe a coach.

This time investigators had a major clue—a witness had seen Christopher and a young male walking toward a tan or cream sedan just prior to the abduction. Under hypnosis the woman described the stranger as being of similar build to the victim and wearing a woolen hat pulled down; she even managed to recall the first seven digits of the car license plate. Unfortunately, the Nebraska Department of Motor Vehicles had registered more than a thousand cars beginning with those seven numbers. Tracing them all would be an arduous task.

Worldwide Search

Meanwhile, analysis of the rope continued. Every manufacturer in the United States was checked, as well as government and military outlets. Abroad, FBI field offices conducted inquiries—particularly in the Far East, where much of the world's rope is made—but every lead fizzled out.

On January 11, 1984, Barbara Weaver, who ran a Bellevue day care center, noticed a young man in a car slowly cruising the street outside. Three times he drove past. Alerted by the recent publicity, she made a note of the car's license plate. The driver saw her do this and stopped. In a fury he demanded that she give him the piece of paper. When Weaver refused, he first threatened to kill her, then roared off. Within seconds she was phoning the police.

Through this license number, detectives traced the vehicle to a local Chevrolet dealership, where they learned that the car was on loan to an enlisted airman at Offutt while his vehicle was being repaired. His own car, a cream-colored sedan, matched the first description exactly, even down to the license plate, which had the same digits recalled by the woman who had undergone hypnosis.

Agents went to Offutt and the quarters of John Joubert, a twenty-one-year-old, boyish-looking, and slightly built radar technician. In appearance and, it was later learned, in history, too, Joubert fit the psychological profile with lethal precision, even down to the fact that he was an assistant scoutmaster of a local troop. In a duffel bag belonging to Joubert, officers found a hunting knife and a length of rope. At first

Joubert denied all knowledge of the crimes, but when confronted with the rope—a very rare kind brought back from Korea by his head scoutmaster—he confessed, saying he was glad to be arrested, otherwise he would have kept on killing.

FBI Laboratory technicians compared the rope found in Joubert's quarters with that used to bind Danny Joe Eberle and declared them microscopically identical. Moreover, a hair found in Joubert's car was found to have come from Danny.

A background check revealed that Joubert, a former altar boy, originated from Portland, Maine, where on August 22, 1982, eleven-year-old Ricky Stetson had been abducted and stabbed to death. Because photographs of the victim's body showed bite marks, impressions of Joubert's teeth were shown to odontologist Dr. Lowell Levine, director of the Forensic Science Unit of the New York State police. He concluded that Joubert's bite marks and those found on the victim were identical. Although he had no known police record, Joubert had a history of attacking young children, each assault worse than the one before.

In July 1984 Joubert pleaded guilty to the double murder in Nebraska and was later sentenced to death. He was later convicted of the murder in Maine. Twelve years of death row appeals ran out at 12:14 A.M. on July 17, 1996. In his final statement, Joubert said, "I just want to say that again I am sorry for what I have done. I do not know if my death will change anything or if it will bring anyone any peace, and I just ask the families of Danny Eberle and Christopher Walden and Richard Stetson to please try to find some peace and ask the people of Nebraska to forgive me. That's all." An hour later, Joubert was strapped into the electric chair and executed.

Conclusion

Ironically, on the very day that Joubert was arrested, the Bellevue authorities received notification from an FBI office in Alabama that yellow fibers in the rope indicated that it had probably been made for the military in the Far East.

Malcolm Fairley

DATE: 1984

LOCATION: Buckinghamshire, England

SIGNIFICANCE: A single flake of yellow paint ended the career of a vicious serial rapist known as the Fox.

Throughout the hot summer of 1984, a hooded burglar subjected residents of a tri-county area north of London to a reign of terror as he broke into houses seemingly at will. Victims who heard his voice described a northern accent; others noticed that he was left-handed. The Criminal Records Office in London provided local police forces with information about all burglars who had migrated from the north of England. This sounded promising, until the computer produced 3,011 names, a daunting number.

On May 12 came a sinister escalation: A man returning home late at night in Cheddington, Buckinghamshire, found himself confronted by a hooded burglar pointing his own shotgun at him. After being tied up, the homeowner was sexually abused; then the robber fled with the shotgun.

As the thefts and attacks increased in both severity and number—the burglar was now using the shotgun to terrify his victims—police saturated the areas where he operated. Nevertheless, the crimes continued unabated. On July 10, the man dubbed the Fox graduated from burglar to brutal rapist. At another Buckinghamshire house, he first tied up a husband, then forced him to watch as he assaulted his wife.

Seventy-two hours later, in Edlesborough, Bedfordshire, three teenagers—a brother and sister and the sister's boyfriend—were watching TV at 2 A.M. when a man wearing a hood and leather gauntlets and brandishing a shotgun burst into the room. He ordered the girl to bind the two boys with electric cord, then dragged her into the bedroom and tied her to the bed with a pillowcase over her head. After helping himself to a drink from the kitchen, the intruder returned to the bedroom and raped her. Afterward, he brought the family dog into the bedroom and attempted to induce an act of bestiality. When the animal failed to respond, he ordered the two boys at gunpoint to have sex with the girl; both simulated the act. The burglar raped the girl again and then sexually assaulted the boys. Afterward, he went downstairs and sat watching videos before leaving.

One month later, he made the mistake that brought about his downfall. On August 17, while driving north to Yorkshire to visit his mother, he was overcome with sexual craving. Outside the village of Brampton, he parked in a quiet field. Without his usual hood, he was forced to improvise one from a pair of green overalls. Then he broke into the home of an accountant—through a bedroom where children were asleep—and tied up the man and his wife. After ransacking the house, he returned to the bedroom and tied the legs of the thirty-seven-year-old housewife apart, one ankle shackled to the bedpost, the other to her husband's leg. When she resisted, he rammed the shotgun into her face and raped her. Afterward, he displayed his security awareness by washing the woman to remove bodily fluids and hair samples; he then cut out a square of bedsheet on which there were semen traces and took it with him.

Blunder

The next day, a police officer searching nearby fields for clues noticed footprints, then tire tracks where a car had been parked. A closer examination revealed a tiny flake of yellow paint on a tree, evidence that someone—possibly the Fox—had backed a car up carelessly, scraping the bodywork. Also found was the square of sheet, the makeshift mask, and a leather glove, all apparently dropped by accident. Of far greater significance, though, was the discovery of the shotgun, carefully concealed beneath a mound of leaves.

Convinced that they were dealing with the Fox, investigators speculated that he might return to pick up the shotgun and so ordered a media blackout of the night's events. To explain their unaccustomed presence in the area, officers staged a fake road accident, while their colleagues installed special infrared cameras and sensitive microphones around where the gun had been hidden in the undergrowth. After dark, a cordon of officers armed with night-sight binoculars encircled the copse. But the Fox did not return.

Nevertheless, he had left several valuable clues. The flake of car paint was identified as Harvest Gold. Only one manufacturer—Austin—used such a color. Further refinement of the search parameters narrowed the field to one model—an Allegro manufactured between May 1973 and August 1975. Because the collision with the branch had occurred at a height of forty-five inches, it was thought that the rapist's car would also be damaged at that height. Now the problem was to identify a northern burglar who drove a scratched yellow Allegro.

Still the main bulk of police efforts were directed toward eliminating those 3,011 burglars known to have migrated south from northern England. On September 11, two policemen sent to check yet another name on the computer printout turned into a north London street and stopped dead in their tracks. Outside the very address they were seeking stood a thin-faced young man washing a bright yellow Austin Allegro. The man, whose name was Malcolm Fairley, invited them into the flat he occupied with his wife and two children. When asked to account for his recent movements, he sounded evasive, so the officers asked if they could examine his car. On the rear bodywork, at a height of approximately forty-five inches, they noticed a chip of paint missing.

Inside, on a seat, lay a man's watch. When one of the officers asked Fairley to put it on, he strapped it to his right wrist—he was left-handed. A search of his flat revealed two sets of green overalls made from material identical to that used for the hood in the Brampton break-in. Fairley was placed under arrest and, faced with the overwhelming body of evidence against him, he quickly confessed.

At the Old Bailey on February 26, 1985, he was sentenced to six terms of life imprisonment.

Conclusion

Had the British police had access to a database of car paint samples, such as that maintained by the FBI, identification of the yellow flake might have been expedited. Even so, that alone would not have led them to Fairley's doorstep. For that, they had to thank the latest computer wizardry, and what is likely to remain the policeman's greatest ally—shoe leather.

VOICEPRINTS

In 1941, the intriguing possibility that someone could be identified solely by the sound of his or her voice led scientists at Bell Telephone Laboratories to develop the sound spectrograph. First used during World War II by the intelligence services to identify voices broadcast by German military communications, the idea fell into disuse until the early sixties, when the FBI approached Bell for assistance in the grouping of voices. Bell engineer Lawrence Kersta became convinced that voice spectrograms, or "voiceprints" as he called them, could provide a valuable means of personal identification, and after four years of research, he branched out on his own, marketing both the principle and the equipment necessary for its implementation.

Kersta's data was based on recordings of fifty thousand different voices, many of them apparently similar. All showed great differences spectrographically. He even employed professional mimics for his trials, asking them to record imitations of individuals and then comparing the results with recordings of the person imitated. The results on graph and screen were again markedly dissimilar.

The spectrograph records a 2.5-second band of speech on high-quality tape and then scans it electronically, a process that takes ninety seconds. The output is then recorded onto a rotating drum. As the drum revolves, a filter adjusts the various frequencies, enabling a stylus to record their intensity. The resulting print contains a pattern of closely spaced lines showing all the audible frequencies in the tape segment. The horizontal axis registers how high or low the voice is at that point; volume is depicted by the pattern density—the denser the print, the louder the tone. Two kinds of voiceprint can be obtained: bar prints and contour prints. Bar prints are used for identification, whereas contour prints are suitable for computerized filing.

Kersta contends that because the parts of the body used in speech—the tongue, teeth, larynx, lungs, and nasal cavity—vary so much in size and shape among individuals, it is virtually impossible for any two people to have identical features. More easily disputed is his claim that a voice remains constant throughout life. Other experts point out that a person's body—and consequently his or her voice—changes with age. Also, a person's way of speaking may change if he or she moves to a different locality and acquires the accent of its people.

Although the rate of voiceprint acceptance in American courtrooms has been dilatory, it has been used as inceptive evidence by the police and has led to the capture of several criminals. In Europe, the discipline is viewed with skepticism. And yet voice individuality has been recognized elsewhere. U.S. Air Force engineers, eager to improve security in restricted areas, have implemented systems whereby in order to gain access a person must satisfy the computer that his or her voice matches a prerecorded sequence in the computer's memory. So far, the system has a claimed success rate of 99 percent, with professional impersonators able to defeat the system just 1 percent of the time. And it is that 1 percent that makes voiceprints so controversial.

Clifford Irving

DATE: 1971

LOCATION: Paradise Island, Bahamas

SIGNIFICANCE: Millions of dollars were at stake in one of the most bizarre crimes on record.

In January 1971, a successful author, Clifford Irving, approached the McGraw-Hill Book Company with a literary bombshell: He had been granted access to write the authorized biography of the world's most famous eccentric—Howard R. Hughes. It was a mouth-watering prospect. For decades, the billionaire industrialist had piqued the public imagination, first as an aviator and moviemaker, more recently as a casino magnate whose fabulous wealth and craving for privacy allowed him to indulge an allegedly outlandish lifestyle.

McGraw-Hill's first inclination was to show Irving the door, but when he produced a clutch of letters all purportedly written by Hughes, and which experts verified as genuine, the company realized it had a potential megaseller on its hands. In anticipation of the windfall to come, it advanced Irving $765,000, an unheard-of sum for those days. Irving duly completed a twelve-hundred-page manuscript. Several people familiar with Hughes read the manuscript and vouched for its authenticity. Prepublication serial rights went to *Time* magazine; everyone stood to make a fortune, or so it seemed.

But then the man at the epicenter of this maelstrom decided that enough was enough. After fifteen years of being incommunicado, Howard Hughes broke his silence to announce that the biography was a fraud—he had never met Irving and had no idea where he had obtained the material for the book, which he dismissed as "totally fantastic fiction."

Hughes imparted these revelations from his Paradise Island sanctuary in the Bahamas via a two-hour teleconference arranged through NBC with selected reporters in a Los Angeles hotel. In typical Howard Hughes fashion, the meeting had been arranged in total secrecy; only later would a transcript be made available to the public. All of the reporters were seasoned veterans who had covered Hughes in his earlier years. Their task was to establish that the ethereal voice at the other end of the telephone line was actually that of Howard Hughes. Twenty minutes were allotted to bombard Hughes with questions about his background in efforts to weed out an impostor. One of the first queries concerned a distinctive cockpit design on one of Hughes's earliest aircraft. The response, cogent and lengthy, revealed a person conversant with all facets of aero-manufacture.

Personal Reminiscences

Other questions were more mundane. "At the time of your round-the-world trip in 1938," asked Marvin Miles of the *Los Angeles Times,* "a superstitious woman placed a good-luck charm on your airplane. What was it and where was it placed?"

This stumped Hughes completely. After a lengthy delay, he said, "I want to be completely honest with you. I don't remember that one." But even before Miles could finish his next question, Hughes jumped in. Yes, it came back to him now—the well-wisher in 1938 had stuck a wad of chewing gum on the tail of his plane. And so it went, every question answered to the reporters' satisfaction, convincing everyone present that they were talking to the genuine Howard Hughes. Relieved to have that part of the proceedings behind him, Hughes then unleashed a tirade against the alleged biography.

"I only wish I was still in the movie business," he said, "because I don't remember any script as wild or as stretching of the imagination as this yarn has turned out to be. . . . I don't know him [Irving]. I never saw him. I have never even heard of him until a matter of days ago when this thing first came to my attention." About one more point he was equally emphatic—not a single penny of the McGraw-Hill advance had wound up in his bank account.

While eager to accommodate their reticent host, the reporters were also straining at the leash to quiz him about his rumored bizarre appearance. Hughes dismissed stories about waist-length hair and eight-inch fingernails as media inventions. In the future, he said, he intended to adopt a more public lifestyle; perhaps then the press would lose interest in him.

When news of the mysterious teleconference was released, it caused pandemonium. Irving naturally denounced the caller as an impostor, describing the voice as "much too vigorous and deep. . . . For anyone who hasn't seen him for twenty-five years it was an excellent forgery." He closed piously: "My obligation, of course, is to Howard Hughes and not the voice on the telephone."

In anticipation of this charge, NBC had already hired voiceprint pioneer Lawrence Kersta to examine the tapes and decide if this was the real Howard Hughes. Because the fundamental principle of voiceprinting hinges on the comparison of one voice with another, obviously another tape of Hughes's voice—one indisputably his—had to be found. It came from a speech made by the billionaire some thirty years earlier to a Senate subcommittee. Kersta knew that this would provide a searching test of his belief that although voices may change over the decades, their inherent individuality remains constant.

Deliberate Analysis

After several days spent converting the tapes into spectrograms; measuring pitch, tone, volume, and other elements of the voices; and then comparing the voice pictures on a line-by-line basis, Kersta declared himself satisfied that the person speaking on the phone was Hughes and no one else. "We are as near to 100 percent positive as a scientist would ever allow himself to be."

Not everyone was convinced, especially those who felt that Kersta's commercial promotion of voiceprinting might well constitute a conflict of interest. Dr. Peter Ladefoged, professor of phonetics at UCLA and a critic of Kersta's early methods, was asked to review the tapes. Ladefoged had to admit that recent technological advances in voiceprinting had removed many of his doubts. After spending hours with a spectrograph, he announced, "I'm reasonably certain that those recordings are of the same voice. . . . Even considering the age difference, . . . it is difficult to believe that this could be two different voices."

From this point on, Irving's scheme began to unravel with almost indecent haste. A Swiss bank, unwilling to be a party to obvious fraud, broke traditional bank secrecy laws and revealed that a check from McGraw-Hill for $650,000 had been deposited in one bank by "H. R. Hughes" and then switched to another account in the name of "Helga R. Hughes"—actually Irving's wife.

On June 16, 1972, found guilty of forgery and forced to repay what was left of the publisher's money, Irving was sentenced to thirty months in prison. He served just over half the jail term before resuming his career as a writer.

Conclusion

Until his death in 1976, Howard Hughes continued to exert an extraordinary hold on the public imagination. To the end, he remained elusive and shadowy. Like his voice, nothing had changed.

Brian Hussong

DATE: 1971

LOCATION: Suamico, Wisconsin

SIGNIFICANCE: A murder of terrifying brutality was solved through the inventive use of voiceprint technology.

Neil LaFave loved his work as a game warden. Long days spent patrolling the Sensiba Wildlife Area near Green Bay, Wisconsin, occasionally enlivened by brushes with poachers, made for a good life, one he wouldn't have swapped with anyone. If there was a special urgency in his step on the afternoon of September 24, 1971, it was understandable. Today was his birthday—he was thirty-two years old—and his wife, Peggy, had planned a party for after work. Neil reckoned it wouldn't take long to post the "closed" signs around this last stretch of waterfowl terrain, plenty of time to be home by six.

By 6:30 P.M., Peggy was eyeing the clock nervously. No sign of Neil. She did her best to placate the guests as they arrived, but there was no disguising her edginess. Eventually, with no sign of the guest of honor, they left, giving Peggy a chance to contact Neil's boss, Harold Shrine. He, too, was puzzled; Neil should have been finished hours ago. Straightaway he drove out to the waterfowl reserve.

All of the "closed" signs were in place, but there was no trace of the young warden. Shrine searched until darkness and then called Peggy. It was time to contact the police. After a fruitless nightlong vigil with powerful searchlights, daylight brought the searchers their first clue—LaFave's green pickup truck, door ajar, empty.

Nearby was an ominous pool of blood, some shattered sunglasses, and two spent .22 caliber shells. Tracing a grisly trail of blood and what looked like human matter, searchers found fragments of bone, and then a tooth, until they reached an area of freshly dug ground. Just a few inches below the surface lay the decapitated body of Neil LaFave. The head, found close by in another shallow grave, had been hacked off with a spade or something similar. Two shots to the skull and several more to the body had ended LaFave's life before the mutilation began.

Such extreme barbarism, police felt, amounted to a clue. This was no random killing. Sergeant Marvin Gerlikovski, the officer in charge of the investigation, aware of LaFave's reputation for tenacity—he had arrested dozens of poachers—suspected that this murder had been motivated by revenge. He ordered that everyone ever arrested by LaFave be tracked down and interviewed. This was no easy task. Many had left the area, and others were reluctant to cooperate, but gradually the number was whittled down. Those unable to satisfy the interrogators were asked to submit to a polygraph examination. With varying degrees of enthusiasm all complied—except one.

Bitter Feud

It was no secret that Brian Hussong hated the dead warden. He and LaFave had been crossing swords for years. No matter how many times he was arrested, the twenty-one-year-old Hussong kept on poaching. Just recently LaFave had caught him again, shooting pheasants. Hussong, sullen and unresponsive, was evasive when questioned about his whereabouts on the day of the murder. He knew his rights, too; when asked if he would undergo a polygraph test, he refused point-blank.

Stymied on this tack, Gerlikovski followed his instincts and applied for a court order to place a wiretap on the house where Hussong lived with his girlfriend, Janice Obey. Sure enough, before long Hussong was overheard discussing guns and alibis with both Janice and his mother, Mary. Even more damning was a conversation with his eighty-three-year-old grandmother, Agnes Hussong, in which she assured him that his guns were hidden where no one could find them.

So stunned was this elderly matriarch when detectives came knocking at her door that she led them straight to the cache of firearms. These were shipped to the state crime laboratory in Madison for testing, where ballistics expert William Rathman confirmed that one of the guns recovered had fired those spent .22 caliber shells found near LaFave's body. On December 16, 1971, nearly three months after the murder, Hussong was taken into custody.

As a precautionary measure, Gerlikovski had recorded all of the phonetap conversations from Hussong's house and passed these tapes to Ernest Nash, head of the Michigan Voice Identification Unit, at that time the nation's most sophisticated voiceprint laboratory.

During testimony, Agnes Hussong bitterly denied ever saying she had hidden her grandson's guns, or that she even had them at all. It was the denial that prosecutor Donald Zuidmulder had both expected and prepared for. He immediately summoned Nash to the stand. Nash first explained the principles behind voiceprinting, then declared himself in no doubt that it was Agnes Hussong's voice on the tape in which she admitted hiding the firearms. All of the other voices on the tapes were similarly identified as belonging to close relatives of Hussong. It was unanswerable.

After three and a half hours of deliberation, during which they asked to hear the tapes again, the jury delivered a guilty verdict. Hussong received life imprisonment. In August 1981, Hussong broke out of Fox Lake Correctional Institution. He remained at large for three months, holed up in a remote cabin, until December 10, when he was killed in a shootout with the police.

Conclusion

This was the first occasion on which Wisconsin had sanctioned the use of wiretaps on the residence of a murder suspect. Given the outcome, it is perhaps surprising that the practice has not become more widespread.

Jimmy Wayne Glenn

DATE: 1973

LOCATION: Modesto, California

SIGNIFICANCE: As a forensic tool, the Psychological Stress Evaluator has been controversial, but few could dispute its usefulness in helping establish the identity of this savage murderer.

All morning Bonnie Johnson had been trying to contact her mother on the phone. But every time, the phone just rang. The more Bonnie tried, the more anxious she became, her mind teeming with all kinds of possibilities, none of them pleasant. Her unease only heightened that afternoon—May 1, 1973—when she knocked at her mother's apartment and heard the same stony silence. Convinced that something was amiss, she obtained a passkey from the manager and let herself into the apartment. Her worst fears were realized with the discovery of Gloria Carpenter's lifeless body submerged in the bathtub. Everything pointed to the fifty-nine-year-old beautician's suffering a heart attack while bathing, a common enough occurrence—until it was noticed that two locks on one door were unfastened. No one with Gloria Carpenter's almost obsessive preoccupation for security would have used the bath without first checking that she was safely locked in.

Not until the arrival of Deputy Coroner William Fanter was the true cause of death known. The thin line around Gloria Carpenter's neck, almost invisible to the naked eye, was clear evidence that she had been strangled, most likely by the nylon stocking found on the bathroom floor. She had also been raped. County pathologist Dr. William Ernoehazy fixed the time of death at about 12:30 that morning. A lack of water in the lungs indicated that she was dead before immersion in the bathtub; the killer's attempt to bury his murderous deeds beneath a patina of natural causes had failed dismally. Because there were no signs of forcible entry, and because nothing was stolen, detectives reasoned that the dead woman had probably known her killer.

Piecing together Gloria Carpenter's whereabouts on the night of her death led investigators to a local bar, where patrons remembered her drinking with a loner named Jimmy Wayne Glenn. When interviewed, Glenn readily admitted carousing with Gloria Carpenter and taking her home, but he denied ever entering the apartment. The evening had ended at her front door, he said. Afterward, he had gone home.

At Glenn's apartment, investigators found a large collection of detective magazines; one, lying on a couch, was opened to a murder story in which the details were eerily similar to the killing of Gloria Carpenter. Glenn insisted that it was mere coincidence, but those listening weren't so sure. They knew that many killers had drawn vicious inspiration from such material. However, despite their suspicions, there wasn't a scrap of concrete evidence to connect Glenn with the murder of Gloria Carpenter. It was at this point that the suggestion was made to use a recently invented but as yet unproved device known as the Psychological Stress Evaluator (PSE).

New Technology

Like the polygraph, the PSE seeks to identify deceit. Unlike the so-called lie detector, it does not track changes in heart rate or perspiration but variations in the voice. It also differs in another important respect. Whereas the polygraph has often been criticized because the attachment of bands and wires to the subject can induce stress and therefore yield faulty results, the PSE bypasses physical contact altogether. Instead, the subject merely speaks into a microphone, or his or her voice is recorded on tape for later evaluation. These sources are then fed into a machine that looks similar to a typewriter. The results are printed out in a graph.

Proponents of the technology claim that stress is reflected in the inaudible variations of the voice, and while these differences cannot necessarily be heard, they can be detected and recorded by the PSE. Once a graph of the subject's voice has been produced, a skilled interpreter can examine it for signs of deception.

Although Glenn would have been quite within his legal rights to decline such a test, he blithely agreed. In the presence of a tape recorder, Glenn was again asked if he had entered Gloria Carpenter's apartment on the crucial night. Again he said no. But when his tape-recorded words were fed into the PSE, the needle swung wildly. The extremes recorded, while enough to convince experts that Glenn was lying, would not be admissible in court, but they did have the effect of steering every investigative effort in one direction—Jimmy Wayne Glenn.

Those efforts were rewarded when a second, more painstaking examination of the bathroom revealed something that had been missed earlier—a faint palm print on the bathtub rim. It matched that of Glenn. Presentation of this evidence to Glenn triggered an abrupt change of story. Yes, he had been in the apartment on the night in question, and other times, as well. And it was on one of those occasions, he said, that he had slipped on the wet bathroom floor and grabbed the tub rim to pull himself upright.

Revisionism of this order received a frosty reception from the jury. They found Glenn guilty of murder, and he was jailed for life.

Conclusion

Although it is unlikely that any court would admit PSE analysis into testimony, its value has been clearly demonstrated on several occasions. Often the accused, when presented with the PSE results, will cut his or her losses and confess.

APPENDIX

Forensic Pioneers and Their Cases

Michael Baden (1939–) This controversial figure's frequent clashes with authority have never overshadowed his technical brilliance. Most recently he has achieved public recognition as a defense witness in the trial of O. J. Simpson.

SIGNIFICANT CASES: Joseph Christopher
Richard Kuklinski

Francis Camps (1905–1972) Crusty and opinionated, Camps was a crusader who sometimes allowed strongly held views to cloud facts. Students gaped at the extraordinary pace of his work in the mortuary and often felt the lash of his tongue if they could not keep up. Tragically, his misjudgments are remembered more than his successes.

SIGNIFICANT CASES: Steven Truscott

John Glaister (1892–1971) A great teacher who succeeded his father as Regius Professor of Forensic Medicine at Glasgow University, Scotland, in the 1930s, together with Sir Sydney Smith and Professor Harvey Littlejohn, Glaister elevated Scottish pathology to a position of international supremacy.

SIGNIFICANT CASES: Buck Ruxton

Edward O. Heinrich (1881–1953) One of the most remarkable figures in the history of U.S. jurisprudence, this Wisconsin-born scientist almost single-handedly created the role of the expert witness. Just the mention of his name was enough to send shudders through opposing counsel, and not for nothing did he become known as "The American Sherlock Holmes."

SIGNIFICANT CASES: The D'Autremont Brothers
Eva Rablen
Charles Schwartz
Jesse Watkins

Milton Helpern (1902–1977) The doyen of American pathologists, Helpern spent decades as chief medical examiner for New York City, a term of service that gave him an unparalleled experience of violent death. It was largely through his untiring efforts that New York City recognized the need to maintain fully equipped forensic facilities.

SIGNIFICANT CASES: Carl Coppolino
Steven Truscott

Alexandre Lacassagne (1844–1921) Early experience in the French military gave Lacassagne the interest in bullet wounds that led him into medicine. First holder of the chair in forensic medicine at the University of Lyons, Lacassagne dealt with all aspects of scientific detection. His dictum, "One must know how to doubt," has become the backbone of sound forensic practice.

SIGNIFICANT CASES: Michel Eyraud

Keith Simpson (1907–1985) Massively experienced, having probably performed more autopsies (more than one hundred thousand) than anyone else, Simpson was a first-rate witness who never lost sight of the fact that juries are largely composed of individuals whose grasp on forensic matters is tenuous at best. His evidence, couched in the clearest of terms, was a model of lucidity and erudition.

SIGNIFICANT CASES: Harry Dobkin
Steven Truscott

Sir Sydney Smith (1883–1969) Although he never achieved the public eminence of his great contemporary, Bernard Spilsbury, Smith took second place to no one in the scope and manner of his work. Always eager to pass on his knowledge to future generations, he can truly be reckoned among the greats of forensic medicine.

SIGNIFICANT CASES: Patrick Brady
Buck Ruxton
Jeannie Donald
Sidney Fox
Patrick Higgins

Sir Bernard Spilsbury (1877–1947) For thirty-five years, the cry of "Call Dr. Bernard Spilsbury" was enough to pack courtrooms throughout England. A man of immense gravitas and seemingly without a shred of self-doubt in his body, he swayed juries and judges alike, earning himself a reputation of near infallibility. Uniquely among the great pathologists, he never published and he never taught. In the end, depressed by his waning powers, he gassed himself.

SIGNIFICANT CASES: Sidney Fox
Patrick Mahon
Norman Thorne
Joseph Williams

Index